On China's Military Rise

Editor

Ming-Hsien Wong

Tamkang University Press

On China's Military Rise

Editor

Ming-Hsien Wong

ISBN: 978-986-5982-85-0

Published in Taiwan by

Tamkang University Press

151, Yingzhuan Rd., Tamsui Dist., New Taipei City 25137, R.O.C.

Tel: 886-2-8631-8661; Fax: 886-2-8631-8660

http://www.tkupress.tku.edu.tw/

E-mail: tkupress@www2.tku.edu.tw

Foreword

In 2009, China replaced Japan as the world's second-largest economy. Since then, the notion of "China is rising" becomes "China rises." In addition, what is rising does not confine to economy but multiple aspects as well. For example, the People's Liberation Army (PLA) has quickly updated their weapons through foreign including foreign procurement, co-manufacture, and indigenous development. The pace of China's strategic ambition surprises everyone, for example, successful test flight of the fighter J-20 and commission of the first carrier Liaoning. To work with the carrier, the fighter J-15 had made successful landed on the carrier. In the ballistic missile, the Dong Feng series, such as DF 21-D, DF 31 and DF 41, have showed improvement on both quality and quantity. Furthermore, after the announcement of establishing the East China Sea Air Defense Identification Zone (ADIZ), China has strengthened the area control capacities in East China Sea. China also extends its influence on South China Sea through islands-building and integrating maritime security institutions. Moreover, China shows and exercises its maritime projection power by humanitarian assistance operations.

The defense budget of China has increased significantly in the past 20 years. Nowadays, its amount is second only to the United States (US). Still, many think tanks believe that the real defense budget of China should be twice or three times than the official figure. As a result, China's defense spending is censored in dark and its strategic intention is doubtful. Although the US department of defense annually publishes a report on the military power of China, the US still can't figure out the China's strategic intention through the development of the PLA.

In fact, after Xi Jinping became the president of China, he proposed many strategies for national security. In addition to the comprehensive "Chinese Dream" and the construction of the Central National Security

Commission (CNSC) for defense integration, he emphasized the "strong-army dream" to support China's development. President Xi expressed that a new commander system is required in order to deal with the new situation and military competition when he met the representatives of the PLA services chief staffs meeting on 22nd September 2014. It means that Beijing plans to establish a joint military command structure in the new era of revolution in military affairs (RMA). In order to increase the capability for local wars under conditions of information technology, China might take reform of its military regions or services to cope with traditional and non-traditional security issues.

Moreover, the Politburo of the Communist Party of China (CPC) adopted the outline of a national security strategy on 23rd January 2015. The new strategy, which is led by president Xi, is formed on the perception that China is facing unpredictable and unprecedented dangers. Therefore, China has to prepare for danger in times of peace. Under the strategy, China will legislate the National Security Law to enforce defense capabilities under the absolute leadership of the CPC's efficient and unified command.

What is the main strategic target of China's military development? Is China merely for national security and preparing for challenges in the future? Or is it paving the way to become a great power? From a history perspective, it generally concludes that the rise and fall of a great power follows a regular circle of expansion, good governance, civil unrest, foreign invasion and the end of an empire. This book collects papers about "China's military rising," which are published in Tamkang School of Strategic Studies 2015 Annual Events. The papers analyze China's military rising from different dimensions, including theory and practice, micro and macro. Followings are the introductions of each paper:

1. Fabrizio Bozzato's "The Red Side of the Moon: China's Pursuit of Lunar Helium"

He thinks that for years, Beijing has been systematically and patiently building up the key competence and platforms needed for an advanced lunar exploration program, under the conviction that 'walking on the Moon' is a reflection of a country's comprehensive national power. China stresses that its space program is for peaceful purposes and maintains its lunar mining would be for the benefit of all humanity.

2. Si Fu Ou's "China's A2AD and Its Geographic Perspective"

The term of A2AD is a Western terminology and its approximation in the Chinese strategic concept is China's active strategic counterattacks on exterior lines (ASCEL) and today China's A2AD capabilities have increased dramatically. China's missile arsenal constitutes severe threats to its neighboring countries, and its surface warships and submarines have frequently penetrated the Ryukyu Islands. Due to its position in the center of the first island chain, Taiwan is of great geostrategic importance.

3. Wei Hwa Chen's "China's Military Power Projection in 2020s: Intentions, Capabilities, and Constraints"

The study suggests that there are several possible scenarios for China's military expansion in a future succession. Initially, such involvement would probably take the form of joint military operation by building aircraft carrier battle groups which have long been considered as major instrument for sea power acquisition. Next, building highly modernized sixth generation war planes, J-20 for instance, would be the next step for the involvement. Lastly, precise configuration of forces will be largely determined by the intensity of possible conflict with neighboring states. ;

4. Ying Yu Lin's "PLA Cyber Warfare & Taiwan's Information Security"

Although modernization of the PLA poses a serious threat to other countries, still can't confront US military force in front of them. So,

Asymmetric warfare is the main thinking in PLA, either Anti–aircraft or "acupuncture warfare" (點穴戰). As the advance of information technology, to the PLA, the combination of cyber-attack and Network-centric warfare (NCW) involves not only software but also hardware. PLA Cyber warfare (網電一體戰) not just operates in cyber space, also a part of information warfare.

5. Taeho Kim's "China's Military Rise and East Asian States: A Regional Security Tour d'Horizon"

This essay argues that China's "anti-access capability"—a U.S.-coined term originally developed for a Taiwan crisis—is equally applicable to other major regional cases such as the Spratly disputes and a North Korean contingency. Furthermore, notwithstanding China's continuous efforts to develop and deploy various types and classes of weapons/platforms, it is the Russian systems and technologies that are most capable and thus likely assigned to the highest mission-critical areas.

6. Tsuneo Nabe Watanabe's "Japan's Security Strategy toward the Rise of China from a Friendship Paradigm to a Mix of Engagement and Hedging"

This paper will explore Japan's strategy toward China within the context of highly complex Japan-US-China trilateral relationship, focusing on the historical legacy and balance of power, as well as the influence of domestic political dynamics. Japan's grand strategy can be said to have been shaped by the pursuit of complex security and economic interests within the Japan-US-China triangle, whose history can be traced back to the 1930s. The paper, though, will limit its scope to the period after 1972, when Japan normalized diplomatic relations with the People's Republic of China, to the present, when Japanese strategic thinking is challenged by the rise of China and the relative decline in US military capabilities and Japan's economic influence.

7. Ja Ian Chong's "Reading the Tea-Leaves of China's Force Modernization: Capabilities, Intentions, and a Perplexed Region"

What perplexes China's neighbors is how Beijing intends to use its expanding military and paramilitary capabilities. Even as Beijing demonstrates great enthusiasm toward using its growing military prowess for the above types of missions, China also appears ready to use apply its newfound military capacities as coercive substitutes for regular diplomacy vis-a-vis other regional actors. This is especially apparent in Chinese approaches to maritime disputes in the East and South China Seas.

8. Richard D. Fisher's "Will It Be Possible To Deter China Into The 2020s?"

China has used its military and economic power to threaten and constrain Taiwan in the 1990s and 2000s and continues to accumulate forces and capabilities for attacking or invading the island democracy. In the current decade China may decide it can use limited force against Japan and the Philippines to enforce maritime area claims, which would damage Washington's regional military leadership while prompting many states to consider strategic deterrent capabilities. Into the 2020s and beyond it will no longer be a question of deterring Chinese military might in Asia alone, but also those Chinese forces that can be deployed globally to influence conflicts that would affect Western or democratic interests.

9. Ming-Shih Shen's "China's Military Capabilities toward Taiwan: Use and Implication"

It is interesting to note that the rapid growth of China's military capabilities in recent years both in the quality and quantity will possess diverse and strong means to coerce Taiwan into negotiation and unification by the military capabilities. But this paper tried to evaluates the PLA's military capabilities form the framework of intention and strategy, to study

PLA how to use these capabilities to be the threats and their implications to Taiwan.

I would like to express my thanks those who have helped to make this book possible. Foremost among those to be acknowledged is the President of Tamkang university, Dr. Flora Chia-I Chang, who greatly supported the publication. Thanks are due as well to the staffs of Tamkang University Press, including Sinn-Cheng Lin, Ciou-Shia Wu, and Yu-Luen Chang for the editing and cover design. Thanks especially due to York W. Chen, the assistant professor in the Graduate Institute of International Affairs and Strategic Studies as the coordinator to make this book possible.

Siou-Jen Chen, the nice and patient assistant in the Graduate Institute of International Affairs and Strategic Studies, helped in administration work. Yu-Jen Jiang, our PhD student, helped to contact all editors. Yin-Sung Su, our MA student, helped to proofread and set type. Without the help from those talents, this book can't be published so quickly. I wish this book can provoke the interests in China's military research and attract more people to get involved. Finally, I wish there could be more and more empirical researches on this issue to achieve the aim of Tamkang School of Strategic Studies, which is "Keep Lead in Taiwan, Become Well-Known Abroad."

Ming-Hsien Wong,

Director of the Graduate Institute of International Affairs and Strategic Studies,

On 26th February 2015,

In T1209, Ching-sheng Memorial Hall, Tamkang University

Contents

The Red Side of the Moon: China's Pursuit of Lunar Helium 3

Fabrizio Bozzato 杜允士

(淡江大學國際事務與戰略研究所博士候選人)

舉頭望明月 *"Upwards the glorious Moon I raise my head."*

[李白 "Li Bai", 701-762]

Abstract

When poets and lovers gaze at the Moon, they might also be looking at a clean and abundant resource that could meet the bulk of the global energy demand. In a world where energy requirements are bound to increase and fossil fuels are finite resources, the cratered satellite may offer mankind a way out of the energy conundrum: Helium-3. This element is a light, non-radioactive, and extremely rare on Earth isotope of helium that is mooted as the fuel of the future to enable nuclear fusion as a power source. It has been calculated that there are over one million tonnes of helium-3 on the lunar surface down to a depth of a few metres. Mining the Moon for the precious isotope, shipping it to our planet - and developing suitable fusion reactors - would provide clean energy for the next millennia. Alas, the costs and scientific challenges of such an enterprise would be phenomenal. Nonetheless, China - which allocates a stable and expectedly growing budget for space activities - appears determined to make it a reality of tomorrow. For years, Beijing has been systematically and patiently building up the key competence and platforms needed for an advanced lunar exploration program, under the conviction that 'walking on the Moon' is a reflection of a country's comprehensive national power. China stresses that its space program is for peaceful purposes and maintains its lunar mining would be for the benefit of all humanity. However, given the absence of wilful competitors, it is also speculated that the Chinese intend to lock up the resources of the Moon and establish a helium-3 monopoly. Thus, the question of whether China will act as a benevolent lunar dragon or create a helium-3 'hydraulic empire' might become one of immense relevance in the next decade.

Keywords:

Moon, China, Helium-3, Nuclear Fusion, Lunar Exploration

Ⅰ. Introduction

"Secure, reliable, affordable, clean and equitable energy supply is fundamental to global economic growth and human development and presents huge challenges for us." [World Energy Council, World Energy Scenarios: Composing energy futures to 2050, 2013]

According to a 2013 United Nations report, the world population is projected to reach 9.6 billion by 2050.[1] Of those future earthlings, 1.6 billion will live in India, and 1.3 billion in China.[2] By then, Nigeria's population is expected to surpass that of the United States.[3] Also, the forty-nine least developed countries are going to double in size from around 900 million people in 2013 to 1.8 billion.[4] In the light of these figures, it is not difficult to understand that humanity is going to face an increasingly acute energy trilemma - how to simultaneously achieve and balance energy security, energy equity (access and affordability) and environmental sustainability - in the coming decades.[5] "Conservatively, [...] more than a nine-fold increase in annual energy production needs to be made available by middle of the

---

[1] "World Population Prospects: The 2012 Revision," *United Nations, Department of Economic and Social Affairs, Population Division, Population Estimates and Projection Section,* February 27, 2014 (last update), <http://esa.un.org/unpd/wpp/index.htm >.

[2] World Bank, "Population Projection Tables by Country and Group," 2014, <http://web.worldbank.org/WBSITE/EXTERNAL/TOPICS/EXTHEALTHNUTRITIONAN DPOPULA-TION/EXTDATASTATISTICSHNP/EXTHNPSTATS/0,,contentMDK:21737699~menuPK:3385623~pagePK:64168445~piPK:64168309~theSitePK:3237118,00.html>.

[3] United Nations, "World Population Prospects: The 2012 Revision".

[4] Ibid.

[5] Christoph Frei et al., *World Energy Scenarios: Composing energy futures to 2050* (London: World Energy Council, 2013), p. 218.

21st Century"[6] and, in the not far future "we have to replace oil, and in the next century we have to replace natural gas - and these two, taken together, represent sixty per cent of the total energy use of every country today."[7] By factoring in that, as the Chinese government's white paper China's Energy Policy 2012 states, "energy is the material basis for the progress of human civilization and an indispensable basic condition for the development of modern society,"[8] it is then easy to see that the trilemma poses a really formidable and frightening challenge.

As the world's second largest energy consumer, China is paying every effort to develop clean and unconventional energy in order to quench its thirst for energy.[9] Beijing is deeply aware of the imperative of addressing the trilemma.[10] In fact, powering an economy the size of China's, especially the size it will be in three decades, only by burning massive quantities of finite fossil fuels and relying on conventional nuclear power is not an option.[11] Besides making China unsustainably energy insecure and growingly

---

[6] Harrison H. Schmitt et al., "Lunar Helium-3 Fusion Resource Distribution," (Madison: University of Wisconsin - Madison, 2011), p. 2.

[7] Guenter Janeschitz as quoted in Raffi Katchadourian, "A Star in a Bottle," New Yorker, March 3, 2014, <http://www.newyorker.com/reporting/2014/03/03/140303fa_fact_khatchadourian>.

[8] Information Office of the State Council of the People's Republic of China, China's Energy Policy 2012, October 2012, p. 2.

[9] "China to Encourage Corporate Participation in Shale Gas Exploration," China Briefing, October 10, 2011, <http://www.china-briefing.com/news/tag/unconventional-energy#sthash.xPnedxzY.dpufand>.

[10] Chang Chung Young and Fabrizio Bozzato, "The Dragon is Thirsty: China's Quest for Energy," paper presented at the conference of "International Conference on the Making of New Asia: Migration, Identity, Interaction and Security" (Taiwan: Fo Guang University, November 5-6, 2011); See also: Jenny Lin, "China's Energy Security Dilemma," Projet 2049 Institute, February 13, 2012.

[11] Joseph P. Giljum, "The Future of China's Energy Security," The Journal of International Policy Solutions, No. 11 (2009), pp. 12-24.

politically unstable, this would eventually result into the country's environ-mental, socio-economic and political collapse, and destructively impact the rest of the world.[12] Also, the rampaging competition for fossil fuels in the international arena would generate intense geopolitical frictions, fuel region-al tensions and breed armed conflicts that would make the international sys-tem savagely Hobbesian and highly flammable.[13] For all these reasons, apart from investing in conventional energy sources, China is also focusing on renewable and unconventional energy, and has made it a strategic priori-ty.[14] Beijing has even officially declared war on pollution,[15] and is not go-ing to leave any stone unturned in the search for a long-term, stable, and bi-osphere-friendly energy source.[16]

China's energy policies are in a state of rapid flux, but coal and other fossil fuels are still the source of the vast majority of China's energy con-sumption today. Currently, Coal accounts for 67 percent of the energy con-sumed in the Asian giant, oil is the second largest source (17 percent).[17] This situation cannot be changed overnight and, as a popular Chinese saying

---

[12] See, for example, Scott Doney, "Oceans of Acid: How Fossil Fuels Could Destroy Marine Ecosystems," *PBS*, February 12, 2014, <http://www.pbs.org/wgbh/nova/next/earth/ocean-acidification/>.

[13] Michael T. Klare, "Fueling the Dragon: China's Strategic Energy Dilemma," *Current History*, Issue 150 (April, 2006), p. 180.

[14] Information Office of the State Council of the People's Republic of China, *China's Energy Policy 2012*, October, 2012.

[15] Mark Ralston, "China to 'declare war' on pollution, premier says," *Reuters*, March 4, 2014, <http://www.reuters.com/article/2014/03/05/us-china-parliament-pollution-idUSBREA240 5W20140305>.

[16] Maria Van Der Hoeven, "Strategizing for Energy Policy: China's Drive to Reduce De-pendence," *Harvard International Review*, Vol. 35, No. 1 (Summer 2013), pp. 14-25.

[17] Joachim Betz, "The Reform of China's Energy Policies," *German Institute of Global and Area Studies Working Papers*, No. 216 (February 2013), p. 6; Also, biomass and waste: 9 percent; hydro-power: 3 percent; Natural gas: 3 percent; nucle-ar power: 1 percent; and other renewable sources: 0.2 percent.

reminds us "water from afar cannot put out a fire close at hand" 遠水救不了近火, id est a slow remedy cannot meet an urgency. For this reason, the Chinese are pouring substantial resources into and placing their bet also on the most futuristic and elusive of unconventional energies: nuclear fusion. In essence, developing nuclear fusion means to develop "what has been labelled 'unconventional nuclear technologies' in order to solve the world's impending energy crisis."[18] Achieving fusion requires sparking and controlling a self-sustaining 'star in a bottle', "using temperatures of 200 million degrees Celsius to get atoms […] to fuse together, releasing huge amounts of energy in the process."[19] Beijing is also actively fostering conventional nuclear power as a source of electricity generation, although it makes up only a very small percentage of generating capacity at present - a fraction that is expected to grow to 6 percent by 2035.[20] However, nuclear fission power plants produce vast quantities of radioactive waste to store, have catastrophic incidents on their record, and are limited by the fact the world's uranium stocks may run out in a couple of hundred years.[21] "Fusion on the other hand," as Steven Cowley - director of the Culham Centre for Fusion Energy and chief executive of the UK Atomic Energy Authority - points out, "gets its fuels, deuterium and lithium, from seawater - not only in plentiful supply but easily accessed, a definite bonus for an increasingly energy-insecure China. Moreover, fusion produces no significant waste. Against the back-

---

[18] Mark Piesing, "Big nuke vs little nuke: how the nuclear establishment is stifling innovation," *Wired*, February 21, 2012,
<http://www.wired.co.uk/news/archive/2012-02/21/nuclear-establishment-hinders>.

[19] Olivia Boyd, "Nuclear fusion: an answer to China's energy problems?" *China Dialogue*, December 2, 2013, <https://www.chinadialogue.net/article/show/single/en/5699>.

[20] Ibid, p. 8.

[21] Steve Fetter, "How long will the world's uranium supplies last?" *Scientific American*, January 26, 2009,
<http://www.scientificamerican.com/article/how-long-will-global-uranium-deposits-last>.

ground of a global struggle to dispose of toxic waste piles, this is a weighty advantage."[22]

Yet, while other scientific challenges have been overcome, a breakthrough in controlled thermonuclear energy (fusion power) seems to be always 'thirty years away'. Notably, The US National Academy of Engineering regards the construction of a commercial thermonuclear reactor, as one of the top engineering challenges of the twenty-first century.[23] Most fusion research has focused on deuterium and/or tritium (heavy isotopes of hydrogen) as fuel for generating fusion. Deuterium is found in abundance in all water on earth while tritium is not found in nature but can be produced by the neutron bombardment of lithium.[24] However, the nuclear fusion Gordian knot could be untied by shifting to another isotope on the periodic table of elements: helium-3.

II . Helium-3: Rare under Heaven

"But when the black gold's in doubt there's none left for you or for me fusing helium-3 our last hope." [Muse, Explorers, The 2nd Law, 2012]

Helium-3 is a light, non-radioactive isotope of helium with two protons and one neutron. Even though this gas is found naturally as a trace component in reservoirs of natural gas and also as a decay product of tritium - one of the elements used in making the hydrogen bomb - there is extremely little

---

[22] Steven Cowley as quoted in Olivia Boyd, "Nuclear fusion: an answer to China's energy problems?"

[23] Raffi Katchadourian, "A Star in a Bottle," *The New Yorker,* March 3, 2014, <http://www.newyorker.com/reporting/2014/03/03/140303fa_fact_khatchadourian?current Page=all>.

[24] Egbert Boeker and Rienk van Grondelle, *Environmental Physics: Sustainable Energy and Climate Change* (Chichester: John Wiley & Sons, 2011), p. 66.

helium-3 on our planet.[25] In 2010, University of Wisconsin-Madison's nuclear chemist Layton J. Wittenberg calculated that the potential helium-3
availability from natural and man-made resources on Earth for scientific experimentation was a mere 161 Kgs.[26] The stockpile of nuclear weapons, the
best current terrestrial source of the gas, provides only a supply of 15 kg cir-
ca a year. Helium-3 has applications in many domains. On the one hand, it is
used in complex low temperature physical measurements as well as in certain magnetic resonance imaging (MRI) in hospitals. On the other hand, the
gas has such valuable military applications that the U.S. army's security services use it for the detection of dirty bombs.[27] Although helium-3 is already
in high demand for many reasons, it could become a universally coveted
commodity thanks to its extraordinary energy properties, namely for its future use in nuclear fusion to generate electric power with no dangerous and
long-lasting radioactive by-products.[28]

Fission power plants use a nuclear reaction to generate heat which turns
water into steam which then hits a turbine to produce power. Current nuclear
power plants have nuclear fission reactors in which uranium nuclei are split
apart. This releases energy, but also radioactivity and spent nuclear fuel that
is reprocessed into uranium, plutonium and radioactive waste which has to

---

[25] Marsha R. D'Souza, Diana M. Otalvaro and Deep Arjun Singh, *Harvesting Helium-3 from the Moon* (Worcester: Polytechnic Institute, 2006), pp. 18-25.

[26] Layton J. Wittenberg, "Helium-3 Resources and Acquisition for Use as Fusion Fuel in Ar- ies III," in Farrokh Najmabadi, Robert W. Conn, et al., *The ARIES-III Tokamak Fusion Re- actor Study - The Final Report*, University of California-San Diego, Advanced Energy Technology Group, Center for Energy Research, p. 15-5.

[27] Dana A. Shea and Daniel Morgan, "The Helium-3 Shortage: Supply, Demand, and Options for Congress," *Congressional Research Service, 7-5700*, December 22, 2010, pp. 1-20.

[28] Satish Kumar and Kopal Gupta, "Helium-3 As An Alternate Fuel Technology (for Produc- ing Electricity)," *Journal of Department of Applied Sciences & Humanities*, Vol. IV (2006), pp. 77-84.

be safely and time-proof stored.[29] For decades scientists have been working to obtain nuclear power from nuclear fusion rather than nuclear fission. In current nuclear fusion reactors, the hydrogen isotopes tritium and deuterium release atomic energy when their nuclei fuse to create helium and a neutron. Nuclear fusion employs the same energy source that fuels the Sun and other stars, without yielding the radioactivity and nuclear waste that is the by-product of nuclear fission power generation.[30] However, the 'fast' neutrons released by nuclear fusion reactors fuelled by tritium and deuterium lead to significant energy loss and are immensely difficult to contain.[31] One potential solution may be to use helium-3 and deuterium - "substituting helium-3 for tritium significantly reduces neutron production, making it safe to locate fusion plants nearer to where power is needed the most, large cities"[32] - or helium-3 alone as the fuel in 'aneutronic' (power without emission of neutrons) fusion reactors. "Perhaps the most promising idea is to fuel a third-generation reactor solely with helium-3, which can directly yield an electric current - no generator required. As much as seventy percent of the

---

[29] World Nuclear Association, "Nuclear Power Reactors," November 2013, <http://www.world-nuclear.org/info/nuclear-fuel-cycle/power-reactors/nuclear-power-reactors/>.

[30] World Nuclear Association, "Nuclear Fusion Power," February 2014, <http://www.world-nuclear.org/info/current-and-future-generation/nuclear-fusion-power/>.

[31] Lawrence Berkeley National Laboratory, "Nuclear Fusion Power," *Guide to the Nuclear Wall Chart,* August 9, 2000 (last updated), <http://www.lbl.gov/abc/wallchart/chapters/14/2.html>.

[32] Stefano Coledan, "Mining The Moon," *Popular Mechanics*, December 7, 2004, <http://www.popularmechanics.com/science/space/moon-mars/1283056>.

energy in the fuels could be captured and put directly to work,"[33] out-pacing coal and natural gas electricity generation by twenty percent.[34]

Nuclear fusion reactors using helium-3 could therefore provide a highly efficient form of nuclear power with virtually no waste and negligible radiation.[35] In the words of Matthew Genge, lecturer at the Faculty of Engineering at the Imperial College in London, "nuclear fusion using Helium-3 would be cleaner, as it doesn't produce any spare neutrons. It should produce vastly more energy than fission reactions without the problem of excessive amounts of radioactive waste."[36] Moreover, eliminating the use of slightly radioactive tritium in the fusion process, by using deuterium and helium-3 for fuel, also has the benefit of simplifying the engineering to meet radiation standards. Also, tritium is not an abundant, naturally occurring isotope of hydrogen on Earth, because of its short half-life of 12.3 years. For Deuterium-Tritium fusion, the tritium would have to be bred from lithium, in a blanket surrounding the inside of the fusion reactor, which is a complication that would be eliminated with Deuterium-Helium-3 fusion.[37]

---

[33] Singam Jayanthu, Bhishm Tripathi and Arjun Sandeep, "Scope of Mining on the Moon - A Critical Appraisal," paper presented at the conference of "Golden Jubilee celebration & MineTECH'11 of The Indian Mining & Engineering Journal" (Raipur: November 18-19, 2011), p. 2.

[34] Keith Veronese, "Could Helium-3 really solve Earth's energy problems?" *io9*, November 5, 2012, <http://io9.com/5908499/could-helium-3-really-solve-earths-energy-problems/all>.

[35] Gary Pajer, Mary Breton, Eric Ham, Joseph Mueller, Micheal Paluszek, A.H. Glasser and Samuel Cohen, "Modular Aneutronic Fusion Engine," Princeton Plasma Physics Laboratory, May 2012.

[36] Matthew Genge as quoted in Henry Gass, "Plans to strip mine the moon may soon be more than just science-fiction," *The Ecologist*, July 4, 2011, <http://www.theecologist.org/News/news_analysis/962678/plans_to_strip_mine_the_moon_may_soon_be_more_than_just_sciencefiction.html>.

[37] Marsha Freeman, "Mining the Moon To Power the Earth," *Executive Intelligence Review*, January 24, 2014, <http://www.larouchepub.com/other/2014/4104moon_power_earth.html>.

Actually, the Helium-3 fusion process is not simply theoretical. The University of Wisconsin-Madison Fusion Technology Institute successfully performed helium-3 fuelled fusion experiments.[38] To date, scientists have only been able to sustain a fusion reaction for a few seconds, but with nothing near the scale or energy yield necessary to be released for commercial use.[39] In fact, the development of commercial fusion reactors is dependent upon demonstrating 'break-even': producing as much energy as it is needed to start the reaction.[40] So far, deuterium-Helium-3 or Helium3-Helium 3 fusion has not yet come close to break-even.[41] However, with massive investments in nuclear fusion research, commercial fusion reactors might become a reality within the next three decades.[42] At that point, the demand for Helium-3 would skyrocket. Presently, even though nuclear fusion does not even work properly yet, helium-3 is so scarce and in demand that in 2010 the US Department of Energy officially lamented a critical shortage in the global supply[43] and is already worth US$16 million per kilo.[44]

---

[38] Matt Treske, "Moon Power," *Wisconsin Engineer*, Vol. 116, No. 1 (November 2011), <http://old.wisconsinengineer.com/articles/191>.

[39] National Academy of Engineering, "Provide energy from fusion," 2012, <http://www.engineeringchallenges.org/cms/8996/9079.aspx>.

[40] Bruno Maffei, "The Physics of Energy sources Nuclear Fusion," University of Manchester, 2012, p.10

[41] Mohammad Mahdavi and Behnaz Kaleji, "Deuterium/helium-3 fusion reactors with lithium seeding," *Plasma Physics and Controlled Fusion*, Vol. 51, No. 8 (July 2009), pp. 85003-0

[42] Sergei V. Ryzhkov, "Alternative Fusion Reactors as Future Commercial Power Plants," *Journal of Plasma and Fusion Research*, Vol. 8 (April 2009), pp. 35-38.

[43] David Kramer, "DOE begins rationing helium-3," *Physics Today*, June 2010, <http://ptonline.aip.org/journals/doc/PHTOAD-ft/vol_63/iss_6/22_1.shtml?bypassSSO=1>.

[44] Henry Gass, "Plans to Strip Mine the Moon May Soon be More Than Just Science-Fiction," *Global Research*, July 7, 2011, <http://www.globalresearch.ca/plans-to-strip-mine-the-moon-may-soon-be-more-than-just-science-fiction/25542?print=1>.

Indeed, Helium-3 is really rare 'under Heaven'. How about 'above Heaven'? Actually, the Sun - like all stars - continuously emits helium-3 within its solar wind, which consists largely of ionized hydrogen and ionized helium. The reason why Helium 3 is so rare on the Earth is that the terrestrial atmosphere and magnetic field prevent any of the solar helium-3 from arriving on our planet. However, as the Moon does not have an atmosphere, there is nothing to stop helium-3 arriving on the surface of our satellite and being absorbed by the lunar soil.[45] Given that The Moon has been bombarded for billions of years by solar wind, helium-3 is available in the dust of the lunar surface.[46] It has been calculated that there are about 1,100,000 metric tonnes of helium-3 on the lunar surface down to a depth of a few metres (since the regolith - i.e. the lunar soil - has been stirred up by collisions with meteorites).[47] More precisely, according to two Chinese scientists, the lunar inventory of Helium-3 is estimated as 6.50×1^{8} kg, where 3.72×1^{8} kg is for the lunar nearside and 2.78×1^{8} kg is for the lunar far side.[48] Helium-3 could potentially be extracted by heating the lunar dust to around 600 degrees C, before bringing it back to the Earth to fuel a new generation of nuclear fusion power plants.[49] Professor Gerald Kulcinski, Director of the Fusion Technology Institute, University of Wisconsin-Madison, maintains that about 40 tonnes of helium-3 - which equate to two fully-loaded Space Shuttle cargo bay's worth - could power the United States for a year at the current

[45] Christopher Barnatt, "Helium-3 Power Generation," *ExplainingTheFuture.com*, September 13, 2012, <http://www.explainingthefuture.com/helium3.html>.

[46] Harrison Schmitt, *Return to the Moon: Exploration, Enterprise, and Energy in the Human Settlement of Space* (New York: Copernicus Books, 2007), pp. 48-51.

[47] "Lunar Helium-3 as an Energy Source, in a nutshell," *Artemis Society International*, 2007, <http://www.asi.org/adb/02/09/he3-intro.html>.

[48] Wenzhe Fa and Ya-Qiu Jin, "Quantitative estimation of helium-3 spatial distribution in the lunar regolith layer," *Icarus*, No. 190 (April 2007), pp. 15-23.

[49] Alfred O. Nier and Dennis J. Schlutter, "Extraction of Helium from Individual IDPs and Lunar Grains by Pulse Heating," *Meteoritics*, Vol. 27, No. 3 (July 1992), < http://adsabs.harvard.edu/abs/1992Metic..27Q.268N >.

rate of energy consumption, without causing smog, acid rain and radioactive waste.[50] This would require mining an areas the size of Washington, D.C. Besides, several other valuable materials - such as oxygen, nitrogen, and carbon monoxide and dioxide - will be produced in the course of recovering the helium-3.[51] It comes as no surprise, then, that the gas has a *potential* economic value in the order of US$ 1bn to 3bn a tonne, making it the only thing remotely economically viable to consider mining from the Moon given current and likely-near-future space travel technologies and capabilities.[52]

III. "Upwards the glorious moon I raise my head"

"Be praised, my Lord, through Sister Moon and the stars; in the heavens you have made them, precious and beautiful." [Francis of Assisi, Canticle of the Sun, 1224]

A team of University of Wisconsin scientists has calculated that if the entire lunar surface were mined, and all of the helium-3 were used for fusion fuel on Earth, it could meet world energy demand for over 10,000 years. In addition, given the estimated potential energy of a ton of helium-3 (the equivalent of about 50 million barrels of crude oil),[53] helium-3 fuelled fusion could free the world from fossil fuel dependency, and is likely to increase mankind's productivity by orders of magnitude.[54] But to supply the planet with fusion power for centuries, humanity has first to return to the Moon. Although mining helium-3 on the cratered satellite to power the Earth

---

[50] Richard Bilder, "A Legal Regime for the Mining of Helium-3 on the Moon: U.S. Policy Options," *Fordham International Law Journal*, Vol. 33, No. 2 (January 2010), p. 246.

[51] Marsha Freeman, "Mining the Moon to Power the Earth."

[52] Christopher Barnatt, "Helium-3 Power Generation."

[53] Joshua E. Keating, "Is There Money In the Moon? Maybe Someday," *Foreign Policy*, June 18, 2012,
<http://www.foreignpolicy.com/articles/2012/06/18/is_there_money_in_the_moon>.

[54] Marsha Freeman, "Mining the Moon to Power the Earth."

has been in the minds of scientists and political deciders since the end of the Apollo program in the early 1970s, to date only China has embarked on a long-term endeavour to achieve such an ambitious goal, having established a satellite-based lunar exploration program called the Chang'e Project (Chang'e is a fairy living on the moon in a Chinese legend) in 2004.[55] The question is: why China? The opinion of Michael C. Zarnstorff, deputy director of research for Princeton Plasma Physics Laboratory, can assist in the quest for the answer. "They [China] need a lot more energy due to their increasing population, and they really want to get rid of the pollution problems they have."[56] If Beijing is able to mine the lunar helium-3 and effectively use it for fusion power, then it could avert China's environmental crisis. In addition, the People's Republic would become a major energy resource player and "offer a clean energy option to countries looking to wean themselves from oil dependency."[57]

Besides having "lots of cash and lots of educated people,"[58] China is graced with a pervasively strategic culture, according to which thorough preparedness and long-term planning are the keys to success.[59] Also, China's one-party political system, in which the leadership of the Chinese Communist Party is the most important factor in determining the future of the country and the generational turnover at the top happens by co-optation

---

[55] Liu Yuanhui, "Reaching the Moon," *China Radio International's English Service*, December 13, 2013, <http://english.cri.cn/7146/2013/12/12/2561s803034.htm>.

[56] Michael C. Zarnstorff as quoted in Brandon Southward, "China's quest for a new energy source heads to space," *CNN Money*, December 20, 2013, <http://features.blogs.fortune.cnn.com/2013/12/20/chinas-quest-for-a-new-energy-source-heads-to-space/>.

[57] Brandon Southward, "China's quest for a new energy source heads to space."

[58] Steven Cowley as quoted in Olivia Boyd, "Nuclear fusion: an answer to China's energy problems?"

[59] Gilbert Rozman, *Chinese Strategic Thought Toward Asia* (New York: Palgrave Macmillan, 2012), pp. 1-6.

once a decade,[60] guarantees that strategic policies are consistently implemented over many years, rather than being reneged on or upturned at every
budget or change of administration as it is often the case with Western democracies.[61] Normally, in the 'State of the Center' long-term plans are ably
enacted by highly selected practical visionaries undisturbed by democracy's
glitches[62] who are acutely aware that addressing the energy trilemma is vital
for regime survival and that conversion to a sustainable world economy is
the only way to go. Moreover, going to the Moon to harvest helium-3 is syn-
ergistically compatible with and reflective of the values and ambitions of
President Xi Jinping's 'Chinese Dream'. The 'Chinese dream' slogan was
launched soon after Xi's inauguration and has quickly become the new national mantra. The expression is used to describe the aspiration of individual
and collective self-improvement in Chinese society and calls for patriotic
unity under a one-party rule. Interestingly, the 'Chinese Dream' vision also
includes a space exploration *élan*,[63] having Mr. Xi emphasized that "the
space dream is an important part of the dream of a strong nation."[64] Indeed,
for a country like China, spacefaring and moonwalking are greatly instrumental to consolidating its legitimacy as *the* rising power. And Many Chi-

---

[60] Xiaowei Zang, *Elite Dualism and Leadership Selection in China*, (London and New York: Routledge, 2004), pp. 147-162.

[61] Walter Wang, "Long Term Planning Puts China on a Different Path," *CleanTechies*, August 16, 2011,
<http://cleantechies.com/2011/08/16/long-term-planning-puts-china-on-a-different-path/>.

[62] Robert Lawrence Kuhn, *How China's Leaders Think: The Inside Story of China's Past, Current and Future Leaders* (Chichester: John Wiley & Sons, 2011), pp. 580-590.

[63] "Reaching for the Moon," *The Economist*, December 21, 2013,
<http://www.economist.com/news/china/21591884-xi-jinping-has-consolidated-power-quic
kly-now-he-showing-it-reaching-moon>.

[64] Xi Jinping as quoted in "Reaching for the Moon."

nese see their space program as the symbol of their once-impoverished nation's ascension to economic and technological primacy.[65]

As Joan Johnson-Freese, a United States Naval War College in Rhode Island professor who researches China's space activities, has pointed out: "China's getting a lot of prestige, which turns into geostrategic influence, from the fact that they are the third country to have manned spaceflight capabilities, [...] that they are going to the moon."[66] Professor Ouyang Ziyuan （歐陽自遠）, the chief scientist of the Chinese Lunar Exploration Program appears to agree. "Lunar exploration is a reflection of a country's comprehensive national power. It is significant for raising our international prestige and increasing our people's cohesion," he told the media. But the Moon could also become an energy cornucopia. Professor Ouyang explained that solar panels would operate far more efficiently on the airless lunar surface and believes that a "belt" of them could "support the whole world" provided the generated electricity is sent back to Earth via lasers or microwaves.[67] Plus, the Moon is "so rich" in helium-3, that this could "solve human beings' energy demand for around 10,000 years at least."[68] In the light of the statements above, it is clear that Beijing's lunar program represents a triple-win venture. Internationally, lunar expeditions "will increase China's political

---

[65] Cole Pfeiffer, "Asia's space race: China looks to dominate the final frontier," *Foreign Policy Today*, December 11, 2013, <http://www.fptoday.org/asias-space-race-china-looks-to-dominate-the-final-frontier/>.

[66] Joan Johnson-Freese as quoted in Chris Buckley, "China blasts off to moon with rover mission," *Seattle Times*, December 2, 2013, <http://seattletimes.com/html/nationworld/2022376926_chinamoonxml.html?syndication=rss>.

[67] Lulu Zhang, "Chief scientist chides narrow view on lunar project," *China.org.cn*, December 17, 2013, <http://china.org.cn/china/2013-12/17/content_30917626.htm>.

[68] Ouyang Ziyuan as quoted in David Shukman, "Why China is fixated on the Moon," *BBC*, November 29, 2013, <http://www.bbc.com/news/25141597 >.

influence in the world."[69] Domestically, 'conquering the Moon' would bolster the consensus for the political leadership and prop up Chinese national pride. Thirdly, on the energy security side, tapping into the Moon has the potential to make China not only energy self-sufficient and secure, but also turn the Chinese into the 'helium-3 Arabs' of the 21st century, especially in case they get to enjoy the position of monopolists. China would then become not only an energy superpower able to fix its social and environmental problems, but also the center of a global helium-3 hydraulic empire.[70] The "spice must flow, and he who controls the spice, controls the universe!" would ideally say Frank Herbert, the author of *Dune*.[71]

Officially China's lunar program has three official main goals. The first is to gain technological skill. Secondly, the Chinese scientists seek to understand the moon's evolution and compare it with Earth.[72] Thirdly, "in terms of the talents, China needs its own intellectual team who can explore the whole lunar and solar system." Additionally, it is acknowledged that the rationale for a long-term program is that "there are many ways humans can use the Moon,"[73] and that Beijing is planning a lunar base.[74] As for the exploitation of the lunar resources, on the one hand the Chinese have repeatedly

---

[69] Ouyang Ziyuan as quoted in Antoaneta Bezlova "China reaps a moon harvest," *Asia Times Online*, October 30, 2007, <http://www.atimes.com/atimes/China/IJ30Ad01.html>.

[70] Karl A. Wittfogel, *Oriental Despotism: A Comparative Study of Total Power* (New York: Random House, 1957).

[71] Brian Herbert, *Dreamer of Dune: The Biography of Frank Herbert* (New York: Tom Doherty Associates, 2003), p. 172.

[72] Julie Sullivan, "Why is China interested in the Moon? Lunar Program Secrets Revealed," *Headlines and Global News*, November 30, 2013, <http://www.hngn.com/articles/18419/20131130/why-china-is-interested-on-the-moon-lunar-program-secrets-revealed.htm>.

[73] Ouyang Ziyuan as quoted in David Shukman, "Why China is fixated on the Moon."

[74] Marsha Freeman, "China Takes Next Step Toward Lunar Industrial Development," *Beijing Review*, No. 9 (February 27, 2014), <http://www.bjreview.com.cn/forum/txt/2014-02/24/content_598245_2.htm>.

declared that they are going to utilize them "to benefit the whole of man-kind,"[75] on the other hand, Professor Ouyang has tellingly remarked that "Whoever first conquers the Moon will benefit first."[76] In order to achieve such goals and eventually 'use the Moon,' the lunar exploration program consists of three stages: 1) *flying around the Moon*. Respectively in 2007 and 2010, Beijing launched the Chang'e-1 and Chang'e-2 unmanned lunar probes to circle the Moon and map its surface to get three-dimensional images of the body from space. Scientists then analyzed the information sent back by the orbiters. 2) *Landing on the moon*. In December 2013 the Chang'e-3 mission, incorporating a robotic lander and China's first lunar rover, reached the Moon. The wheeled rover explored the vicinity of the landing area and radar-scanned the lunar subsurface structure. The second phase of the program will be completed by the Chang'e-4 mission, incorporating a robotic lander and rover, which is scheduled for launch in 2015. 3) *Returning from the moon*. The Chang'e-5 mission may be launched in 2017 or 2018 to further explore the Moon and collect lunar soil, and then would return soil and rock samples to China for first-hand examination. Only after the completion of these three phases China will be finally able to land human beings on the Moon.[77] According to British space scientist Richard Holdaway, China could have astronauts treading on the regolith by 2025.[78]

---

[75] Information Office of the State Council of the People's Republic of China, *China's Space Activities in 2011 - I. Purposes and Principles of Development*, December 29, 2011,<http://china.org.cn/government/whitepaper/2011-12/29/content_24280462.htm>.

[76] Ouyang Ziyuan as quoted in Ajey Lele, *Asian Space Race: Rhetoric Or Reality?* (London: Springer India, 2013), p. 170.

[77] Ling Xin, "An Interview with Ouyang Ziyuan: Chang'e-3 and China's Lunar Missions," *Bulletin of the Chinese Academy of Sciences*, vol. 27, no. 4 (November 2013), pp. 26-31.

[78] David Shukman, "Why China is fixated on the Moon."

IV. To the Moon (and beyond)!

"We choose to go to the moon. We choose to go to the moon in this decade and do the other things, not because they are easy, but because they are hard, because that goal will serve to organize and measure the best of our energies and skills, because that challenge is one that we are willing to accept, one we are unwilling to postpone, and one which we intend to win, and the others, too." [John F. Kennedy, Moon Speech at Rice University, 1962]

As observed by former US astronaut and geologist Harrison Schmitt, Chinese scientists and experts have framed Beijing's space program partially in terms of their nation's constant quest for energy and raw materials, "talking about -helium-3 and solar power as potential energy sources on the moon, as well as its reserves of titanium, rare earths, uranium and thorite."[79] This 'pursuit of lunar resources' theme has then been combined with a 'geopolitical competition' discourse conveying a sense of urgency. "If China doesn't explore the moon, we will have no say in international lunar exploration and can't safeguard our proper rights and interests,"[80] Professor Ouyang declared in 2010, hinting that progress in the lunar program would confer an edge to China if and when the extraction of the Moon's riches turns political. The 15 December 2013 edition of the *Beijing Youth Daily* argued that "China can obtain a certificate to sharing lunar interests only by carrying out exploration and gaining actual results." It also contended: "How to protect China's

---

[79] Harrison Schmitt as quoted in Simon Denyer "China launches 'Jade Rabbit' rover to moon, precursor to manned mission," *Washington Post*, December 2, 2013, <http://www.washingtonpost.com/world/china-launches-jade-rabbit-rover-to-moon-precursor-to-manned-mission/2013/12/02/87ba7d1a-5b13-11e3-801f-1f90bf692c9b_story.html>.

[80] Ouyang Ziyuan as quoted in Jonathan Adams, "Dragon watch: China pulls ahead in moon race," *Global Post*, November 2, 2010, <http://www.globalpost.com/dispatch/china/101027/space-race-moon?page=0,1>.

interest in outer space has become an inevitable question."[81] Dean Cheng, an expert on China's space program at the Heritage Foundation, got the message clear. "Once you start mining, and even before, questions arise as to ownership, as to profit-sharing (if any), as to who has the ability to establish and enforce claims in space," he said. "A long-term presence in space will give China political capital."[82] Thirdly, the lunar program has been presented to policy makers and the general public as a cost effective investment. For example, the chief scientist of the program has stressed that the total spending of Chang'e-1 mission was about RMB 1.4 billion, the same amount as the money used to construct two kilometers of subway in Beijing.[83] Similarly, the Chinese political leaders can be reminded that, according to experts in the US , the total estimated cost for fusion development, rocket development and starting lunar operations would be about US$15-20 billion over two decades.[84] By comparison, another big nuclear fusion project (on Earth), the International Thermonuclear Reactor Project (ITER) has an estimated total cost of now €15 billion (US$20.5 billion),[85] and going to the Moon to mine helium-3 would cost "about the same as was required for the

---

[81] Beijing Youth Daily "ANALYSIS: Lunar success marks China's rise as next space power," *The Asahi Shimbun*, December 16, 2013, <http://ajw.asahi.com/article/asia/china/AJ201312160060>.

[82] Dean Cheng as quoted in Jonathan Adams, "Dragon watch: China pulls ahead in moon race."

[83] Ling Xin, "An Interview with Ouyang Ziyuan: Chang'e-3 and China's Lunar Missions," p. 229.

[84] Steve Almasy, "Could the moon provide clean energy for Earth?" *CNN*, July 21, 2011, <http://edition.cnn.com/2011/TECH/innovation/07/21/mining.moon.helium3/>.

[85] Dave Keating, "Oettinger aims to get ITER back on track," *European Voice*, September 5, 2013, <http://www.europeanvoice.com/article/imported/oettinger-aims-to-get-iter-back-on-track/78100.aspx>.

1970s Trans Alaska Pipeline."[86] Actually, US$ 15-20 billion does not appear to be an excessive financial commitment for a country which is to spend US$ 1.7 trillion between 2011-2015 - in the form of investment, assistance for state-owned enterprises, and bank loans - for a plan aiming at covering 11.4 percent of China's energy needs by 2015, and 15 percent by 2020, from non-fossil energy.[87]

Finally, the seductiveness of China's lunar vision has been enhanced with two additional charms: China's technological advancement and solar system exploration. As for the first, Ouyang Ziyuan's speeches often mention the achievements of the U.S. Apollo program (1963-1972) in order to illustrate the transformational characteristics of any lunar project. The Chinese scientist reminds his audiences that Washington spent US$ 25.4 billion on the Moon's exploration at that time, which has thus far yielded an output worth fourteen times the original investment, leading to the birth of several new hi-tech industries and technologies such as the rocket, radar, radio guidance and so on, which were then put into civil use.[88] The implication is that China's Moon exploration and colonization are going to be the catalyst for revolutionary technological progress that can transform the country's entire industrial landscape and bring a galaxy of economic and social benefits.[89] However, helium-3 remains the biggest gem on the selenitic crown. If it is postulated that the commercial value of helium-3 will be US$3 bil-

---

[86] Harrison Smith as quoted in Cecilia Jamasmie, "Mining the Moon is Closer than Ever," *Mining.com*, January 1, 2010,
<http://www.mining.com/mining-the-moon-is-closer-than-ever/>.

[87] Arthur Guschin, "China's Renewable Energy Opportunity," *Diplomat*, April 3, 2014,
<http://thediplomat.com/2014/04/chinas-renewable-energy-opportunity/>.

[88] Lulu Zhang, "Chief scientist chides narrow view on lunar project."

[89] South Central University for Nationalities, "Chief Scientist of CLEP Ouyang Ziyuan was Named Honor Professor of SCUN," November 14, 2012,
<http://en.scuec.edu.cn/s/148/t/499/a9/5a/info43354.htm>.

lion/ton,[90] and defensively estimated that there are 1 million tons of the precious gas trapped in the regolith,[91] then the whole stock of lunar helium-3 would be worth an astonishing three quadrillion dollars. That is more than enough to cover for the costs and risks of extracting and shipping it back to Earth. Finally, it should be kept in mind that while rocket fuels and consumables now cost an average of $20,000 per pound to lift off Earth, resources could instead be carried off the Moon much more economically. Given that the lunar gravitational pull is inferior to the Earth's, 83.3% (or 5/6) less to be exact,[92] transporting material from the moon requires just 1/14th to 1/20th of the fuel needed to lift material up from the terrestrial surface.[93]

Financial considerations apart, helium-3 would be crucial for what perhaps is the most ambitious goal of China's lunar program: setting up a lunar base and using the Moon as a stepping-stone for space exploration.[94] In order turn the Moon into an operational headquarters for scientific experimentation and further exploration of the solar system, a lunar base should be established first.[95] Helium-3 would be crucial for achieving that. In fact, the

---

[90] Steve Taranovich, "Helium-3 and Lunar power for Earth reactors," *EDN Network*, March 15, 2013,
<http://edn.com/electronics-blogs/powersource/4410034/Helium-3-and-Lunar-power-for-Earth-reactors>.

[91] University of Wisconsin-Madison - Fusion Technology Institute, "Lunar Mining of Helium-3," March 12, 2014 (updated), <http://fti.neep.wisc.edu/research/he3>.

[92] Fabrizio Tamburini, C.Coufano, M. Della Valle and R. Gilmozzi, "No quantum gravity signature from the farthest quasars," *Astronomy & Astrophysics*, Vol. 533, A71 (September 2011), p. 5.

[93] Ray Villard, "Strip Mine the Moon to Fuel Space Exploration," *Discovery Communications*, July 13, 2011,
<http://news.discovery.com/space/moon-mining-needed-to-fuel-space-exploration-110713.htm>.

[94] Defang Kong and Qian Zhang, "Manned lunar landing under research," *People's Daily Online*, January 8, 2014, <http://english.peopledaily.com.cn/202936/8506408.html>.

[95] Sarah Fecht, "Six Reasons NASA Should Build a Research Base on the Moon," *National Geographic*, December 20, 2013,

immediately available by-products of helium-3 production include hydrogen, water, and compounds of nitrogen and carbon. Oxygen can be easily produced by electrolysis of water. Thus, by mining Helium-3 Moon settlers would be able to obtain the air and water they would need to make lunar colonization sustainable.[96] In essence, extracting helium-3 produces the resources we need to gather more of it.[97] Lunar helium-3 could also become the premier rocket fuel of the future, turning the Moon into the launching pad or a refuelling service station for space-bound missions. It appears that in the permanently darkened craters in the Moon's polar regions there are significant reserves of water (ice) that can be melted, purified and electrolyzed into hydrogen and oxygen.[98] One by product would be hydrogen peroxide for rocket fuel. Hydrogen can be obtained also as a by-product of helium-3 mining. Yet, helium-3 would offer ginormous advantages over hydrogen if it is used a nuclear rocket fuel. As John Slough, a University of Washington's professor of aeronautics and astronautics explains, "Using existing rocket fuels, it is nearly impossible for humans to explore much be-

---

<http://news.nationalgeographic.com/news/2013/12/131220-lunar-research-base-mars-mission-science/>.

[96] William A. Ambrose, James F. Reilly II and Douglas C. Peters (eds.), *AAPG Memoir 101: Energy Resources for Human Settlement in the Solar System and Earth's Future in Space* (Tulsa: American Association of Petroleum Geologists, 2013), p. 41.

[97] At the University of Wisconsin, Dr. Kulcinski and his colleagues have designed a ten ton regolith mining machine called the Mark 3. They predict that one of their Mark 3 robotic miners could process six million tons of regolith per year and produce 201 tons of hydrogen, 109 tons of water, 0.033 tons of helium 3 (that's 33 kg.), 102 tons of helium 4, 16.5 tons of nitrogen, 63 tons of carbon monoxide, 56 tons of CO_2 and 53 tons of methane. The CO, CO_2 and CH_4 contain a total of 82 tons of carbon. These researchers have chosen to heat the regolith only up to 700 C. See: Gerald L. Kulcinski, A Resource Assessment and Extraction of Lunar 3He, Presented at the US-USSR Workshop on D-3He Reactor Studies, September 25 - October 2, 1991, Moscow.

[98] National Space Agency, "Researchers Estimate Ice Content of Crater at Moon's South Pole," June 20, 2012, <http://www.nasa.gov/mission_pages/LRO/news/crater-ice.html>.

yond Earth."[99] NASA estimates a round-trip human expedition to Mars would take more than four years using current technology, but according to Slough the same voyage could be completed in maximum-three-month expeditions on a spaceship powered by fusion.[100] Helium-3 would then be the best candidate as fuel for the fusion engines because it is abundant on the Moon and would provide far more power per unit of mass than chemical rocket fuels.[101] Moreover, helium-3 as fusion fuel greatly reduces neutron production and therefore would be the safest option for the ship crews.[102]

In the light of all these elements, it is clear that China is not just re-enacting and repeating the past of US space program, but intends to shape the future. Beijing's grand plan to mine lunar helium-3 should be understood as "the first step toward creating a scientific and economic revolution which will power global economic growth and open the entire Solar System to mankind."[103]

V. Game of Moons

"A trader from Quarth told me that dragons come from the Moon." [Game of Thrones, Season 1, Episode 2, 2011]

Two years before the Apollo 11 mission, a treaty was signed by the United States, the United Kingdom, and the Soviet Union. Inked even as the

---

[99] John Slough as quoted in Michelle Ma, "Rocket powered by nuclear fusion could send humans to Mars," *University of Washington News and Information*, April 4, 2013, <http://www.washington.edu/news/2013/04/04/rocket-powered-by-nuclear-fusion-could-send-humans-to-mars/>.

[100] Michelle Ma, "Rocket powered by nuclear fusion could send humans to Mars."

[101] John F. Santarius, "Lunar 3He, Fusion Propulsion, and Space Development," Proceedings of the Second Conference on Lunar Bases and Space Activities of the 21st Century (Texas: Houston, Apr 5-7, 1988), p. 75.

[102] John F. Santarius, *Role of Advanced-Fuel and, Innovative Concept Fusion in the Nuclear Renaissance*, APS Division of Plasma Physics Meeting, Philadelphia, October 31, 2006.

[103] Marsha Freeman, "Mining the Moon to Power the Earth."

race to plant a flag on the lunar soil was well underway, the 1967 Outer Space Treaty stipulated that no nation-state could ever own the Moon.[104] According to the Treaty - to which 102 countries, including the People's Republic of China (which joined in 1983), are currently parties[105] - Activities on the Moon may be pursued freely without any discrimination of any kind, and countries can place vehicles, personnel, stations, and facilities anywhere on or below the surface. However, as said above, neither the surface nor the subsurface of the Moon can become the property of any country or its citizens. In fact, a 2009 Statement by the Board of Directors of the International Institute of Space Law clarifies that: "The current international legal regime is binding both on States and, through the precise wording of Article VI of the Outer Space Treaty of 1967, [...] also on non-governmental entities, i.e. individuals, legal persons and private companies. [...]Since there is no territorial jurisdiction in outer space or on celestial bodies, there can be no private ownership of parts thereof, as this would presuppose the existence of a territorial sovereign competent to confer such titles of ownership."[106] Also, as leading international law of space scholar Harold Bashor points out: "There are no rights of ownership for any natural resources in place. [...] This is generally interpreted to mean that a country may not claim ownership of any resources until they have been extracted. Yet, any extraction is required to be for the benefit of mankind according to the Common Heritage of Mankind principle."[107] Moreover, since the Moon is to be explored and

---

[104] K.R., "Lunar property rights - Hard cheese," *Economist*, February 16, 2014, <http://www.economist.com/blogs/babbage/2014/02/lunar-property-rights>.

[105] United Nations Office for Disarmament Affairs, *Treaty on Principles Governing the Activities of States in the Exploration and Use of Outer Space, including the Moon and Other Celestial Bodies*, April 11, 2014, <http://disarmament.un.org/treaties/t/outer_space>.

[106] International Institute of Space Law, *Statement of the Board of Directors of the International Institute of Space Law (IISL)*, March 22, 2009.

[107] Harold Bashor as quoted in Leonard David, "Moon Water: A Trickle of Data and a Flood of Questions," *Space.com*, March 6, 2006, <http://www.space.com/2120-moon-water-trickle-data-flood-questions.html>.

exploited for peaceful purposes, Bashor argues, states have an obligation not to interfere with the activities of any other state on the Moon, and any conflict has to be reported to the United Nations.[108] Alas, the current system is predicated heavily on good faith, and whether future Moon-settling countries will behave fairly is yet to be seen. "A system lacking a clear legal framework has thus far worked for scientific ventures, such as the International Space Station. But history tells a different story when big businesses and competing nations turn their sights on a new frontier."[109]

This said, which states are actually going to play 'game of Moons'? As many as eleven robotic lunar missions including orbiters, rovers and sample return missions are to be launched between now and 2020.[110] China (2015; 2017-18),[111] Russia (2016;2019),[112] India (2017),[113] Japan (2018)[114] and

---

[108] Harold Bashor as quoted in Leonard David, "Moon Water: A Trickle of Data and a Flood of Questions."

[109] Joshua Philipp, "Mining the Moon: Plans Taking Off, but Rules Lacking," *Epoch Times*, January 29, 2014, <http://www.theepochtimes.com/n3/476806-mining-the-moon-plans-taking-off-but-rules-lacking/>.

[110] Craig Covault, "The New Race for the Moon," *SpaceRef*, October 4, 2013, <http://spaceref.com/asia/the-new-race-for-the-moon.html>.

[111] Ningzhu Zhu, "China plans to launch Chang'e 5 in 2017," *Xinhua*, <http://news.xinhuanet.com/english/china/2013-12/16/c_132971252.htm>.

[112] Igor Mitrofanov, Vladimir Dolgopolov, Viktor Khartov, Alexandr Lukjanchikov, Vlad Tret'yakov and Lev Zelenyi, *'Luna-Glob' and 'Luna-Resurs': science goals, payload and status* (Vienna: European Geosciences Union General Assembly 2014, April 27- May 02, 2014).

[113] "India to launch Chandrayaan-II by 2017," *The Hindu*, January 10, 2014, <http://www.thehindu.com/sci-tech/science/india-to-launch-chandrayaanii-by-2017/article5562361.ece?ref=sliderNews>.

[114] Kenichi Fujita, "An Overview of Japan's Planetary Probe Mission Planning," paper presented at the conference of "9th International Planetary Probe Workshop" (Toulouse: JAXA, June 2012).

the US (2018)[115] all plan missions during this period while new players eyeing post 2020 Moon missions may include South Korea (2020),[116] while the European Space Agency's Lunar Lander project has been shelved due to budgetary constraints.[117] Those states giving serious study to the launch of manned Moon missions by 2020-2030 are China,[118], and Russia,[119] Japan (in collaboration with the US),[120] and perhaps India.[121] Even though it appears that there are going to be several contenders playing the selenitic game, what is really needed in order to extract helium-3 and the other lunar resources is a lunar base. Hence, only two chess-pieces remain standing on the board: the People's Republic of China and the Russian Federation. Officials from both Beijing and Moscow have declared that their countries are going

---

[115] National Aeronautics and Space Administration, "ILN - International Lunar Network," April 30, 2013, <http://science.nasa.gov/missions/iln/>; See also: Irene Klotz, "NASA Planning for Mission To Mine Water on the Moon," January 28, 2014

[116] Soo Bin Park, "South Korea reveals Moon-lander plans," *Nature*, November 13, 2013, <http://www.nature.com/news/south-korea-reveals-moon-lander-plans-1.14159>.

[117] Stephen Clark, "ESA lunar lander shelved ahead of budget conference," *Astronomy Now*, November 8, 2012, <http://www.astronomynow.com/news/n1211/20moonlander/#.U0kPxVWSw_k>.

[118] Shaoting Ji and Wen Wang, "China's space exploration goals before 2020," *China Daily*, March 10, 2014, <http://usa.chinadaily.com.cn/china/2014-03/10/content_17336950.htm>.

[119] William Stewart, "Is Vlad keen on a trip? Putin eyes up cosmonaut uniform as his deputy premier sets out plans to colonise space and declares 'We are coming to the Moon FOREVER," *Daily Mail*, April 11, 2014, <http://www.dailymail.co.uk/news/article-2602291/We-coming-Moon-FOREVER-Russia-sets-plans-conquer-colonise-space-including-permanent-manned-moon-base.html#ixzz2yfcYJrJG>.

[120] Srinivas Laxman, "Japan SELENE-2 Lunar Mission Planned For 2017," *Asian Scientist*, July 16, 2012, <http://www.asianscientist.com/topnews/japan-announces-selene-2-lunar-mission-2017/>.

[121] "ISRO: No Manned Mission to Moon," *Express News Service*, January 1, 2014, <http://www.newindianexpress.com/states/karnataka/ISRO-No-Manned-Mission-to-Moon/2014/01/01/article1976540.ece#.U0kjN1WSw_k>.

to build a base on the Moon. On 8 January 2014, Zhang Yuhua - Deputy General Director and Deputy General Designer of the Chang'e-3 probe system revealed that: "In addition to manned lunar landing technology, we are also working on the construction of a lunar base, which will be used for new energy development and living space expansion." [122] His words echo Oyuang Zyuan's 2002 mantic statement: "China will establish a base on the moon as we did in the South Pole and the North Pole."[123] In Russia's case, the announcement has come for the upper echelons. Deputy premier Dmitry Rogozin, who is in overall charge of Russia's space and defence industries, writing about the "colonisation of the moon and near-moon space" on the 10 April 2014 issue of the official daily *Rossiiskaya Gazeta* stated that Moscow plans to establish a lunar base for long-term missions to the Moon by 2040. Rogozin affirmed that that Earth's satellite is the only realistic source to obtain water, minerals and other resources for future space missions. A lunar laboratory complex will also be used for testing new space technologies. "This process has the beginning, but has no end. We are coming to the Moon forever," he promised.[124] Despite Rogozin's rethoric, it might be surmised that the Chinese are better positioned in the race for the lunar helium-3. For a start, they have more money and resources, began "from a long way back but now they are catching up fast,"[125] and "by the end of the decade [...] want to move from being what is classed as a major space power to being a strong space power." Tellingly, "within a little more than a decade, the only work-

---

[122] Zhang Yuhua as quoted in Defang Kong and Qian Zhang, "Manned lunar landing under research," *People's Daily Online*, January 8, 2014, <http://english.peopledaily.com.cn/202936/8506408.html>.

[123] "2010 moon mission for China," *CNN*, May 20, 2002, <http://edition.cnn.com/2002/TECH/space/05/20/china.space/index.html?_s=PM:TECH>.

[124] Dmitry Rogozin as quoted in *The Voice of Russia*, "Russia plans to get a foothold in the Moon - Dmitriy Rogozin," April 11, 2014, <http://voiceofrussia.com/news/2014_04_11/Russia-plans-to-get-a-foothold-in-the-Moon-Dmitriy-Rogozin-5452/>.

[125] Richard Holdaway as quoted in David Shukman, "Why China is fixated on the Moon."

ing space station in orbit could be Chinese."[126] In sum, Beijing backs words with facts. For this reason, when asked if the idea of a Chinese lunar base extracting minerals was remotely plausible, the afore-mentioned Prof. Richard Holdaway: "It is perfectly plausible from the technical point of view, absolutely plausible from the finance point of view because they have great buying power."[127] Buying power will be certainly needed, given that a 2009 analysis by the Center for Strategic and International Studies estimated that a four-person research station on the lunar surface would cost US$35 billion to build and US$7.35 billion per year to operate.[128]

If China wins the 'race for the Moon' and establishes a man-tended outpost conducting helium-3 mining operations, it would create a scenario similar that of the 2009 movie *Moon*. In that motion picture, a private company called Lunar Industries has built a mining base on the Moon and enjoys a helium-3 extraction and shipping monopoly - the same kind of monopoly that in the past created the fortunes of ventures like the East India Companies.[129] Unlike that fictional universe, in the case of a Chinese lunar base the monopoly would be held by a state. The ramifications and consequences of such a scenario would be 'cosmic'. First, "China is what international relations scholars call a 'revisionist power,' seeking opportunities to assert its

---

[126] Kevin Pollpeter as quoted in Sarah Cruddas, "Will China have an Apollo moment?" *BBC*, December 11, 2013, <http://www.bbc.com/future/story/20131211-will-china-have-an-apollo-moment>.

[127] Richard Holdaway as quoted in David Shukman, "Why China is fixated on the Moon."

[128] Vincent G. Sabathier, Johannes Weppler and Ashley Bander, "Costs of an International Lunar Base," *Center for Strategic and International Studies*, September 24, 2009, < https://csis.org/publication/costs-international-lunar-base>.

[129] Duane Byrge, "Firm Review: Moon", *The Hollywood Reporter*, January 26, 2009, <http://www.pastdeadline.com/hr/film-reviews/film-review-moon-1003934260.story>.

enhanced relative position in international affairs."[130] Thus, establishing an automated or manned helium-3 operation on the Moon would be a spectacular statement of *grandeur*.[131] Secondly, due to the inevitable depletion of fossil fuels on Earth, Beijing would be in a position to gradually build a helium-3 hydraulic empire in which it would control the supply of the precious gas, and become the only energy superpower. The making of such an empire would be most likely met with resistance. Plausibly, the prospect of China's energy supremacy, which would undoubtedly transubstantiate into pervasive geopolitical influence, would cause geopolitical tension, agglutinate anti-Chinese alliances, and prompt the other space-faring nations - the US *in primis* - to rush to the Moon to break the Dragon's monopoly. Then, a scenario similar to that described in *Limit*, a science and political fiction novel by Frank Schatzing set in a 2025, seeing China and the US develop a new Cold War for lunar helium-3 taking the space race of yesteryear to new heights. Thirdly, China might decide to acquire or retain control over helium-3 deposits by annexing lunar regions. Signally, international law would be neither an impassable hurdle nor an effective deterrent. Although the 1967 Outer Space Treaty asserts common ownership over everything in the universe beyond the Earth and requires all countries to share in the benefits of space,[132] its article 17 permits signatory states to withdraw from the treaty with only a year's notice.[133] Unilateral withdrawal by one of the major

---

[130] John Hickman, "Red Moon Rising: Could China's lunar ambitions scramble politics here on Earth?" *Foreign Policy*, June 18, 2012, <http://www.foreignpolicy.com/articles/2012/06/18/red_moon_rising>.

[131] John M. Logdson, "Lost in Space," *Politico Magazine*, December 19, 2013, <http://www.politico.com/magazine/story/2013/12/china-moon-landing-us-space-race-10 1278.html#ixzz2ylcNGIYe>.

[132] Everett C. Dolman, *Astropolitik: Classical Geopolitics in the Space Age* (London and Portland: Frank Cass Publishers, 2002), pp. 84-88.

[133] John Hickman, "Still crazy after four decades: The case for withdrawing from the 1967 Outer Space Treaty," *Space Review*, September 24, 2007, <http://www.thespacereview.com/article/960/1>.

spacefaring powers would undermine the existing international legal regime in space, prompting the other players to secure a piece of the pie in the sky for themselves. This would start a period of colonialism reminiscent of that in 19th century. Having established a permanent manned lunar base, China would be able to substantiate its claim by satisfying an important criterion for sovereignty: the wishes of the inhabitants. Also, claims over lunar areas beyond China's 'red side of the Moon' by other powers would legitimize Beijing's acquisition of its new selenitic dominions (where Chinese sovereignty would provide regulations and protection for private investors to operate).[134] Once in control of vast helium-3 fields, China could even astutely play 'game of Moons' by favouring the settlement and encouraging the territorial claims of non-hostile or friendly powers - for example other BRICS countries - in order to contain Western expansion and access to helium-3 on the lunar surface. Finally, China could decide to use its lunar base as a military asset "to dominate access on and off our planet Earth and determine who will extract valuable resources from the moon in the years ahead."[135] More piercingly put, "the Moon could hypothetically be used as a military battle station and ballistic missiles could be launched against any military target on Earth"[136] or in space. Our planet's celestial sister could also become one of the battlefields of future 'helium-3 conflicts', which would be

---

[134] Henry Hertzfeld, "The Moon is a Land without Sovereignty: Will it be a Business Friendly Environment?" *High Frontier Journal*, Vol. 3, No. 2 (Spring 2007), p. 43.

[135] Richard C. Cook, "Militarization and the Moon-Mars Program: Another Wrong Turn in Space?" *Global Research*, January 22, 2007,
<http://www.globalresearch.ca/militarization-and-the-moon-mars-program-another-wrong-turn-in-space/4554>.

[136] *Want China Times*, "PLA dreams of turning moon into Death Star, says expert," December 12, 2013,
<http://www.wantchinatimes.com/news-subclass-cnt.aspx?cid=1101&MainCatID=11&id=20131203000106>.

simultaneously fought or spill-over on lunar, space and Earth domains.[137] If this will turn into 'tomorrow's truth', then helium-3 will not just fuel the future, but also future rivalries and wars. The price for global energy security would then be global geopolitical insecurity.

VI. Conclusion: A Call to Cooperation

"There was a time when energy was a dirty world - when turning on your lights was a hard choice. Cities in brown out, food shortages, cars burning fuel to run. But that was the past, where are we now? How did we make the world so much better, make deserts bloom? Right now we're the largest producer of fusion energy in the world. The energy of the sun, trapped in rock, harvested by machine from the far side of the moon. Today we deliver enough clean burning Helium-3 to supply the energy needs of nearly 70% of the planet. Who'd have thought, all the energy we ever needed, right above our heads. The power of the moon. The power of our future." ['Lunar Industries Commercial' in Duncan Jones, Moon, 2009]

The 'game of Moons' scenarios evoked in the previous pages are not anticipations of an inescapable future. On the contrary, lunar exploration and resources development can be international cooperation synergizers and confidence building catalysts. Consistently, the Beijing Declaration, issued at the 2008 Global Space Development Summit in Beijing, calls for international cooperation "in all the applicative fields of space [...] as the world enters a challenging period characterized by globalization, dramatic population growth, serious environmental concerns and scarcity of resources."[138] By 2050 there will be a dire paucity of all the economically recoverable fossil fuels (there would still be plenty of coal, but can the humankind afford to put up with the greenhouse gases?). "Also, all alternative sources of energy,

---

[137] "How the Moon could fuel World Wars," Metro, May 26, 2009, <http://metro.co.uk/2009/05/26/how-the-moon-could-fuel-world-wars-149344/>.

[138] Beijing Declaration, Global Space Development Summit, Beijing, 24 April 2008, p. 2.

like water power, solar power, tidal power, wind power, geothermal power, and wood will not be sufficient to supply more than 10 percent of the energy which will be needed by the 20 billion people that will be on earth at that time. We will be out of energy and forced to seek a new source,"[139] predicted a venerable scholar at the turn of the millennium. And Sister Moon, "precious and beautiful,"[140] can tend the Earth its energy salvation. The helium-3 trapped into the lunar soil offers humanity about ten times the energy that could be obtained from mining all the fossil fuels on Earth, without causing apocalyptic pollution. Also by tossing all Earth's uranium into liquid metal fast breeder reactors, we could generate about half this much energy.[141] But some men will have to cross the sky and conquer the Moon, and other people will have to tame particles, to open a new future up to humanity. Indeed, the quest for helium-3 is involved with the dynamics of succumbing to or reversing the process of global collapse. Common destiny and enlightened self-interest both dictate cooperation among all space-faring nations. Two countries in particular have greater responsibilities than the others: the US and China.

In almost every area of space activity, the US has a clear technological and operational advantage over the other countries, including China. For example, in 2012 NASA landed the Curiosity rover on Mars, a much more difficult task than the Chang'e 3 mission by any measure.[142] However, the US star does not shine as bright as in the past due to budget cuts, and a re-

---

[139] Wilson Greatbatch "War is not the Answer, Nuclear Fusion Power with Helium 3 is the Answer," *Prometheus*, No. 87, Special Issue (2003), <http://www.meaus.com/greatbatch-war-not-answer.htm>.

[140] Francis of Assisi, "The Canticle of Brother Sun," *The Franciscans*, <http://www.franciscanfriarstor.com/archive/stfrancis/stf_canticle_of_the_sun.htm>.

[141] Satish Kumar and Kopal Gupta, "Helium-3 As An Alternate Fuel Technology (for Producing Electricity)," p. 80.

[142] Keith Cowing, "Is China Really Winning a Space Race with Us?" *NASA Watch*, December 20, 2013, <http://nasawatch.com/archives/2013/12/frank-wolf-wave.html>.

luctance to maintain its space leadership[143] as revealed by the cancellation of the American project designed to take humans back to the Moon (Constellation Program).[144] On the other hand, even though Beijing's overall budget in space programmes is still rather moderate compared with that of the US, China appears to have what Confucius would describe as "the will to win, the desire to succeed, the urge to reach its full potential" in lunar and space exploration. The Chinese are quickly developing their own space technology kung fu and are currently collaborating with other countries such as Russia, Brazil, France, Germany and, very fruitfully, with the European Space Agency[145] - but not with the US. Actually, China has recently made several overtures to the US. For example, Xu Dazhe, the new chief of China's space industry, while attending the International Space Exploration Forum in Washington in January 2014 said: "We are willing to cooperate with all the countries in the world, including the United States and developing countries."[146] "The US however, is wary of entering in any type of collaborative interaction with the Chinese, primarily for national security reasons ranging from technology transfer concerns to a general mistrust of the People's Liberation Army's involvement in Beijing's space program. Consequently, in 2011 Washington "has enacted Public Law 112-10, Public Law 101-246 and Public Law 106-391 to suspend all bilateral activities between

---

[143] Lamont Colucci, "America Must Retake Lead in Space Exploration," *US News*, December 11, 2012,
<http://www.usnews.com/opinion/blogs/world-report/2012/12/11/america-must-retake-lead-in-space-exploration>.

[144] Jonathan Amos, "Obama cancels Moon return project," *BBC*, February 1, 2010,
<http://news.bbc.co.uk/2/hi/science/nature/8489097.stm>.

[145] Jane Qiu, "Head of China's space science reaches out," *Nature*, March 6, 2014,
<http://www.nature.com/news/head-of-china-s-space-science-reaches-out-1.14797>.

[146] Xu Dazhe as quoted in PTI, "China wants space collaboration with US," *Economic Times*, January 11, 2014,
<http://economictimes.indiatimes.com/articleshow/28678995.cms?utm_source=contentofinterest&utm_medium=text&utm_campaign=cppst>.

NASA and the Chinese in spaceflight projects."[147] Furthermore, China was barred from participating in the current orbiting space station, largely because of US objections over political differences.[148] By contrast, the Chinese said they will welcome foreign astronauts aboard their future space station, which is scheduled to become operational in 2020.[149]

While the US government's duty and prerogative to protect national security is not in question, the issue of collaborating with China in space activities should be considered in the light of the benefits of going back to the Moon and establishing a settlement for the production of the Helium-3 fusion fuel. Working with the Chinese as part of a global effort to solve the energy conundrum would then become "the next logical step."[150] For sure, combining forces would make humanity's pursuit of helium-3 power, quicker, cheaper and more efficient. Starting a cooperative effort, inclusive of China and the US, for lunar exploration would, first of all, require each participant a change of mind-set as well as adopting an approach based on the four principles indicated by the Beijing Declaration: mutual benefit, transparency, reciprocity, and cost sharing.[151] Actually, the same document identifies the development of a lunar base as the ideal next project for interna-

---

[147] Sanford Healey, "The Future of United States-Chinese Space Relations," paper presented at the conference of "National Conference on Undergraduate Research" (LA: University of Wisconsin-La Crosse, April 11-13, 2013), p. 342.

[148] Peter Rakobowchuk, "Hadfield: The future of Canadian space exploration lies with China," *Canadian Press*, December 28, 2013, <http://globalnews.ca/news/1052624/hadfield-the-future-of-canadian-space-exploration-lies-with-china/>.

[149] Leonard David, "China Invites Foreign Astronauts to Fly On Future Space Station," *Space.com*, September 28, 2013, <http://www.space.com/22984-china-space-station-foreign-astronauts.html>.

[150] Chris Hadfield as quoted in Peter Rakobowchuk, "Hadfield: The future of Canadian space exploration lies with China."

[151] Beijing Declaration, p. 2.

tional collaboration on space exploration.[152] Creative politics and diplomacy will also play a crucial role in ensuring good governance and fair dividends to all parties. New legal regimes for exploiting helium-3 and other lunar resources could be designed and approved. A new international regime, organization or enterprise for the cooperative development and terrestrial fusion of lunar helium-3 may be needed.[153] Many diverse solutions will be possible as long as a sense of common destiny will be shared by the moon-settling nations. The race for making available a safe, clean and revolutionary source of energy to all human beings should not have any loser, only winners. Thus, civilizational or national egoisms should be left back on Earth. Helium-3 power is not meant to be the flame casting deep shadows over a new Dark Age, but the glorious light of a global renaissance: an era in which people will look at the Moon through a clear unpolluted sky. In Washington as in Moscow, New Delhi or Beijing.

---

[152] Vincent G. Sabathier, Johannes Weppler and Ashley Bander, "Costs of an International Lunar Base."

[153] Richard Bilder, "A Legal Regime for the Mining of Helium-3 on the Moon: U.S. Policy Options," *Fordham International Law Journal*, pp. 289-299.

China's A2AD and Its Geographic Perspective

Si Fu Ou

(Research Fellow, Institute of National Defense and Strategy Studies, and Adjunct Assistant Professor, Graduate Institute of Future Studies, Tamkang University)

Abstract

Washington's and Beijing's access and anti-access strategies are constructed on the basis of two island chains. In order to countervail a rising China, the essence of the U.S. Asia pivot is reviving again the geopolitics of the Western Pacific. The term of A2AD is a Western terminology and its approximation in the Chinese strategic concept is China's active strategic counterattacks on exterior lines (ASCEL) and today China's A2AD capabilities have increased dramatically. China's missile arsenal constitutes severe threats to its neighboring countries, and its surface warships and submarines have frequently penetrated the Ryukyu Islands. Due to its position in the center of the first island chain, Taiwan is of great geostrategic importance. It is the cork that keeps the Chinese naval and air forces bottled up within the China Sea. One simple option for the U.S. and its partner nations to counter Beijing's increasing military strength is to boost their own version of A2AD capability. Taiwan should reshape its force structure with a strategy of asymmetry, combat credibility and resilience, as well as cooperating with the U.S. Asia pivot.

Key words

A2AD, ASCEL, ASBM, island chains, Miyako Strait

Ⅰ. Introduction

When the People's Liberation Army (PLA) launched threatening war games off Taiwan nearly two decades ago on the eve of a presidential election on this island, the U.S. deployed two aircraft carriers, and China quickly backed down. Things don't seem so one-sided any more. While the U.S. military has been drained by over a decade of costly conflicts in Afghanistan and Iraq, China has developed air, naval and missile capabilities that could undercut U.S. superiority in the Western Pacific. Under the name of anti-access and area denial (A2AD), the PLA's growing array of aircraft, naval and submarine vessels, ballistic and cruise missiles, anti-satellite and cyber war capabilities already enable it to project power beyond its shore. It plans new submarines, larger naval destroyers and transport aircraft that could expand that reach further.

This shift raises questions about whether the U.S. can meet its commitment to maintain a strong presence in the Asia-Pacific for decades—not just a matter of global prestige but one also seen as critical for safeguarding shipping lanes vital for world trade and protecting allies. With its military buildup, China poses an unprecedented threat to Taiwan. China regards Taiwan as part of its territory. While relations between the two, long seen as a potential flash point, have warmed in the past several years, China threat is still Taiwan's biggest security challenge. China's assertion of territorial claims in the East/South China Seas, which it has declared as core interests, has spooked its neighbors and fortified their support for a strong U.S. presence in the region. Even former enemy Vietnam is forging military ties with the U.S. As the PLA has gotten more capable and Beijing has behaved more aggressively, a number of countries are looking at the U.S. as a hedge to make sure they can maintain independence, security and stability.

Amid the rivalry between Washington's and Beijing's access and anti-access goals in the Western Pacific island chains, Taiwan can play a role due to its position in the center of the first island chain. Geography does not determine the strategic ambitions or policies of a state, but it does condition the choices made by policy makers, presenting both opportunities and constraints.[1] Geography still matters to Chinese strategists in the twenty-first century. At key moments in the history of China's interaction with other hegemonic powers, Taiwan appears as either a potential bridgehead from which foreign rivals of China may establish themselves close to China's shores and adversely affect China's security or a buffer in China's hands that interposes a territorial layer of security on China's southeastern coast, a symbolic or genuine source of defense against western adversaries. From a geographic perspective, this paper explores Beijing's A2AD contents, capabilities, and threats, and their implications for Taiwan and other neighboring countries in the region.

II. Geopolitical Competition for the Western Pacific

Washington's joint access[2] or gaining and maintaining access[3] and Beijing's A2AD are constructed on the basis of two island chains. In order to countervail a rising China, the essence of the U.S. Asia pivot is reviving again the geopolitics of the Western Pacific.

Classical geopolitics paced emphasis on position and derived power—Halford J. Mackinder on the Eurasian heartland and related land power, Nicholas J. Spykman on the Asian Rimland, and Alfred T. Mahan on the Pacific Ocean and related sea power.[4] Technology has collapsed distance, but

---

[1] Colin S. Gray and Geoffrey Sloan, eds., *Geopolitics, Geography and Strategy* (London: Frank Cass, 1999), p. 2.

[2] U.S. Department of Defense, *Joint Operational Access Concept*, 2012.

[3] U.S. Department of the Army and the Department of the Navy, *Gaining and Maintaining Access: An Army-Marine Corps Concept*, 2012.

[4] Colin Flint, *Introduction to Geopolitics*, 2nd edition (Abingdon and New York: Routledge, 2012), pp. 6-11.

it has hardly negated geography. Rather, it has increased the preciousness of disputed territory. As for the strategic challenge posed to the world by China, Robert Kaplan argues, "we would do well not to focus too single-mindedly on economics and politics. Geography provides a wider lens. The map leads us to the right sorts of questions." [5]

Perhaps the most enduring asymmetry in the Sino-American relation-ship is geographic. The U.S. is an insular power of continental size. It is sur-rounding by oceans that provide open access to the sea and create significant barriers to distant enemies, and it shares land borders with neighbors that are weak, friendly, or both. China, by contrast, lives in a tough and complicated neighborhood. It borders 14 different countries and its land boundaries have 22,117 km. Moreover, due to its vast size and the location of its frontiers, its outlying territories are home to minority groups alienated from Chinese Han majority.[6] Finally, China is hemmed in across its maritime periphery by is-land barriers that it does not control and maritime chokepoints that can be monitored and defended by potential opponents.[7]

For the U.S., isolation does have its drawbacks, including the difficulty of projecting military power overseas across extended lines of communica-tion, as well as the need to sustain a robust forward military presence so that allies abroad remain confident in American security commitments. Never-theless, isolation also confers crucial benefits. Most importantly, the U.S. is far more secure than most other nations, it is the preferred security partner for countries that fear proximate rivals more than a remote hegemon, and its

---

[5] Robert D. Kaplan, "The Revenge of Geography," *Foreign Policy*, May/June 2009, <http://www.foreignpolicy.com>.

[6] Central Intelligence Agency, *The World Factbook: China*, 2013, <http://www.cia.gov/library/publications/the-world-factbook/geos/ch.html>.

[7] Robert D. Kaplan, "Geography Strikes Back," *Wall Street Journal*, September 7, 2012, <http://online.wsj.com/article/SB10000872396390443819404577635332556005436.html? mod=WSJASIA_hpp_MIDDLEFourthNews#printMode>.

access to the sea underpins its naval power and dominance of the maritime commons.[8]

Likewise, China's geography yields a number of advantages. It can sustain military operations along relatively secure interior lines of communication and exploit its strategic depth by locating sensitive targets deep inland. Lord Nelson once joked that "a ship's a fool to fight a fort." It also benefits from greater magazine depth; that is, it can stockpile far more weapons and munitions than a rival operating its forces far from home. Yet geography has been disadvantageous to China in a number of ways. In particular, the need to keep a wary eye on threatening neighbors and restive minorities has compelled Beijing to remain focused on territorial defense and internal security, and to sustain large ground and paramilitary forces. This, in turn, has inhibited China from extending its influence globally. Meanwhile, island barriers constitute significant obstacles that constrain China's access to the Pacific Ocean and Indian Ocean.[9] The advantages and disadvantages of maritime-continental competition are shown in the following Table 1.

---

[8] Evan Braden Montgomery, "Competitive Strategies against Continental Powers: The Geopolitics of Sino-Indian-American Relations," *Journal of Strategic Studies*, Vol. 36, No. 1 (2013), p. 79.

[9] Montgomery, "Competitive Strategies against Continental Powers: The Geopolitics of Sino-Indian-American Relations," p. 80.

Table 1: Geography and the Sino-American Strategic Competition

	China	The United States
Relative Geographic Advantages	• Interior lines of communication • Strategic depth • Magazine depth	• Territorial security due to geographic isolation • less threatening to other nations than nearby rivals • open access to the sea
Relative Geographic Disadvantages	• Surrounded by land-power counterweights • Persistent threat of internal unrest in outlying territories • Constrained access to the sea	• Exterior lines of communication • Forward military presence necessary to reassure distant allies and deter adversaries

Sources: Evan Braden Montgomery, "Competitive Strategies against Continental Powers: The Geopolitics of Sino-Indian-American Relations," *Journal of Strategic Studies*, 2013, p. 79.

From the Chinese geostrategic perspective, China's coastline is quite extensive, but its land-sea orientation was powerfully influenced by the special circumstances of its neighbors; for a time, the sea was viewed as a solid barrier and so was neglected. In modern times, the sea became a springboard for foreign invaders as the great powers smashed in China's maritime gate.

According to Chinese geostrategist Xu Qi, China is part of what Halford Mackinder termed the Inner or Marginal Crescent on the fringe of the Eurasian landmass, with undoubted geostrategic preponderance on the continent. China's sea areas are linked from south to south and connected to the world's oceans; however, passage in and out of the open ocean is obstructed by two island chains. China's maritime geostrategic posture is thus in a semi-enclosed condition. China's heartland faces the sea; the benefits of economic development are increasingly dependent on the sea, and security threats come from the sea. The U.S. has deployed strong forces in the Western Pacific and has formed a system of military bases on the first and second Island chains with a strategic posture involving Japan and South Korea as the northern anchors, Australia and the Philippines as the southern anchors, and Guam positioned as the forward base.[10]

At present, offshore defense is the fundamental guarantee of national maritime security. In the 1970s, Deng Xiaoping promulgated China's strategy of preparation for combat in the offshore area, since the main scope of maritime strategic defense was close in to shore. The distinguishing feature of the maritime strategy put forward on this offshore defense foundation is the realization of national unification, giving a prominent position to the safeguarding of maritime rights and interests, and emphasizing that the navy must be able to respond to a regional war at sea, as well as to neutralize enemy encroachment. As a result, the scope of naval strategic defense should progressively expand. In the direction of the South China Sea, the sea area extends 1,600 nautical miles from mainland China, but the scope of naval strategic defense is still within the first island chain.[11]

Open ocean-area defense is an essential shield for long-term national interests. In the future, some maritime powers may employ long-range strike weapons to attack into the depths of China. The vast, unobstructed character of the naval battlefield is favorable for military force concentration, mobility, force projection, and initiating sudden attacks. Future at-sea informationalized warfare has characteristics of noncontact and nonlinearity, and in particular uses advanced informationalized weapons, space weapons, and new-concept weapons, etc. It can involve multidimensional precision attacks in the sea areas beyond the first island chain and threaten important political,

---

[10] Xu Qi, "Maritime Geostrategy and the Development of the Chinese Navy in the Early Twenty-First Century," trans. Andrew S. Erickson and Lyle J. Goldstein, *Naval War College Review*, Vol. 59, No. 4 (2006), pp. 56-58.

[11] Xu Qi, "Maritime Geostrategy and the Development of the Chinese Navy in the Early Twenty-First Century," pp. 60-61.

economic, and military targets within strategic depth. The maritime security threat comes from the open ocean. This requires the PLA Navy (PLAN) to cast the field vision of its strategic defense to the open ocean and to develop attack capabilities for battle operations on exterior lines, in order to hold up the necessary shield for the long-term development of national interests.[12]

Admiral Liu Huaqing, head of the PLAN from 1982 to 1986, saw control of the waters within its boundaries as the first step in a three-stage strategy to transform the navy into a formidable platform for projecting Chinese power. The next stage, he wrote, involved controlling a second island chain linking the Ogasawara Islands—including Iwo Jima—with Guam and Indonesia, while the third stage focused on ending American dominance throughout the Pacific and Indian oceans, largely by deploying aircraft carriers in the region.[13] As China builds aircraft carriers, a senior PLAN officer argues, China and the U.S. can make a deal. The U.S. takes Hawaii West and China will take Hawaii East and the Indian Ocean.[14]

For China, its projection of power is constrained by geography. It is restricted in the South by the Strait of Malacca and the Association of Southeast Asian Nations (ASEAN); in the North, by the Strait of Korea/Tsushima and Japan and South Korea; and on the East by Taiwan, "the unsinkable air-

---

[12] Xu Qi, "Maritime Geostrategy and the Development of the Chinese Navy in the Early Twenty-First Century," p. 61.

[13] "Disputed Asian Islands Part of First Island Chain that Could Restrict China's Navy," *Washington Post*, December 24, 2012, <http://www.washingtonpost.com/world/asia_pacific/disputed-asian-islands-parts-of-first-island-chain-that-could-restrict-china-navy/2012/12/24/550d3a62-4d89-11e2-835b-02f92c0daa43_print.html>.

[14] Manu Pubby, "China Proposed Division of Pacific, Indian Ocean Region, We Declined: US Admiral," *Indian Express*, May 15, 2009, <http://www.indianexpress.com/news/china-proposed-division-of-pacific-indian-ocean-regions-we-declined-us-admiral/459851>.

craft carrier," as Douglas MacArthur called it. There are three barriers encircling and thwarting China:

1. The arc from Japan to South Korea—Diego Garcia in the Indian Ocean, forming a zone of forward bases.

2. The arc from Guam to Australia.

3. The arc from Hawaii-Midway-Aleutian Islands to Alaska.[15]

On the contrary, the U. S. policy in the Pacific continues to be based on Alfred T. Mahan's perceptions: forward operation bases, positioning assets around chokepoints and SLOCs (Sea Lines of Communication), deploying a navy presence on all seas, and maintaining the capability to intervene at key geostrategic points. For a maritime power, the maritime frontier is, as observed by Homer Lea, one of its enemies.[16] Tanguy Struye De Swielande explains further, this translates into a triple line of defense:

1. Japan-South Korea-Taiwan-Thailand-Singapore.

2. Japan-Guam-Philippines-Australia.

3. Alaska/Aleutian Islands-Hawaii-Samoa.[17]

Homer Lea, who is today best known as close advisor to Dr. Sun Yat-Sen during the 1911 Chinese Republican revolution,[18] insisted on the need to rely on forward operation bases in the form of a triangle. Strategic geometry was the key principle on which much of his thought was based. By forming numerous triangles with Guam as the potential center or node, the

---

[15] Toshi Yoshihara, "Chinese Missile Strategy and the U.S. Naval Presence in Japan," *Naval War College Review*, Vol. 63, No. 3 (2010), p. 43.

[16] Homer. Lea, The day of the Saxon, reprinted (San Diego, California: Simon, 2003).

[17] Tanguy Struye De Swielande, "The Reassertion of the United States in the Asia-Pacific Region," *Parameters*, Vol. 42, No. 1 (2012), p. 83.

[18] Lawrence. Kaplan, *Homer Lea: American Soldier of Fortune* (Lexington, Kentucky: University Press of Kentucky, 2010).

U.S. is actually executing the vision presented by Lea. Some examples are the Guam-Japan-South Korea, Guam-Darwin-Pearl Harbor, and Guam-Taiwan-Japan triangles (Figure 1).[19] Moreover, the U.S. strategic security presence at Guam is underpinned by a wider strategic triangle under direct American control, in the shape of Hawaii, Alaska (including Aleutian islands) and Guam, with Guam playing a particularly important forward apex role for the Western Pacific in the second island chain.[20]

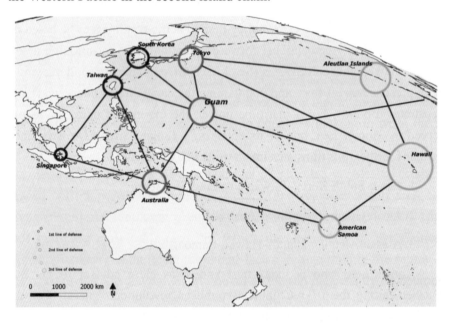

Figure 1: The Triangles in the Pacific Ocean

Sources: Tanguy Struye De Swielande, "The Reassertion of the United States in the Asia-Pacific Region," *Parameters*, Spring 2012, p. 84.

---

[19] Tanguy Struye De Swielande, "The Reassertion of the United States in the Asia-Pacific Region," p. 84.

[20] U.S. Air Force, "Pacific Air forces," *Fact Sheet*, March 24, 2013, <http://www.109aw.ang.af.mil/resources/factsheets/factsheet_p..>.

In the Indian Ocean, the Diego Garcia atoll could fulfill the same purpose as Guam in the Pacific. In October 2002, the U.S. stationed in Djibouti
the Combined Joint Task Force—Horn of Africa (CJTF-HOA) to fight terrorism and monitor sea lanes.[21] In January 2009, the U.S. established Joint
Task Force (JTF) 151 to fight piracy.[22] This is a spin-off from the Navy
CJTF-150 that was created in 2002 to fight terrorism. The U.S. also has the
5th Fleet (with headquarters in Bahrain) with smaller units and organizations
throughout the Gulf States. Like a number of other nations, the U.S. is interested in the island of Socotra (Yemen). Washington has also reinforced its
presence in Kenya (Manda Bay and Lamu). In order to dominate the Indian
Ocean, there is a necessity to control the Diego Garcia-Seychelles-Mauritius
triangle, strategically located between Asia and Africa. Although Diego Garcia is the only location with American bases on it, there is continuing cooperation between the Seychelles and the U.S. in the fight against piracy. The
Seychelles hosts a number of U.S. drones used for monitoring piracy activities in the Indian Ocean. These drones also carry out strike missions against
Al Qaeda affiliates in Somalia.[23] Even in light of these ongoing American
initiatives, China and India continue their influence over the Seychelles and
Mauritius. New Deldi, which already has perceived these island nations as
India's sphere of influence, would not want to be outflanked by China in the
region.[24]

Apart from the tensions in the Sino-American rivalries, if there is a serious prospect of an extended security competition with China over the skies

---

[21] "10 Things about CJTF-HOA," *U.S. Africa Command Blog*, June 1, 2012,
<http://africom.wordpress.com/2012/06/01/10-thing-about-cjtf..>.

[22] Jennifer Cragg, "US Navy, Partners Deter Pirate Attacks," *The Pentagon Brief*, January 31,
2009, <http://pentagonbrief.wordpress.com/tag/jtf-151>.

[23] Zenko and Welch, "Where the Drones are," *Foreign Policy,* 2012,
<http://www.foreignpolicy.com/articles/2012/05/29/where_the_drones_are?page=full>.

[24] Pathak, "China and Francophone Western Indian Ocean Region: Implications for Indian
Interests," *Journal of Defense Studies,* Vol. 3, No. 4 (2009), p. 89.

and waters of the western Pacific, Taiwan will be pivotal in shaping how it unfolds. Partly this is the consequence of geography. Taiwan is smack in the middle of the first island chain. Tetsuo Kotani conceded that "losing Taiwan to the PLA would be a game changer for Japan and the regional naval balance."[25] James Holmes argues further that during World War II, Admiral Ernest King aptly portrayed Formosa as the cork in the bottle of the South China Sea—as a base from which naval and air forces could seal off Imperial Japan's Southern Resources Area. Similarly, Taiwan has long served as a literal and figurative cork in China's bottle, riveting Beijing's attention on the cross-strait stalemate while complicating north-south movement along the Asian seaboard and access to the Western Pacific.[26]

Taiwan is important geopolitically. However, facing a rising China, there is a containing or accommodating China debate in the U.S. It translates a supporting or abandoning Taiwan policy dispute in other words. Former Vice Chairman of the Joint Chiefs of Staff, Admiral Bill Owens, suggests the Taiwan Relations Acts (TRA) is outdated.[27] Zbigniew Brzezinski calls Taiwan an endangered species.[28] Bruce Gilley describes Taiwan's current tra-

[25] Vance Serchuk, "Obama's Silence on Taiwan Masks Its Significance in Asia," *Washington Post*, May 24, 2013, <http://www.washingtonpost.com/opinions/obama-silence-on-taiwan-masks-its-significane-in-us-relations-with-china/2013/05/23/a1b40470-c243-11e2-914f-a7aba60512a7_print.html>.

[26] James Holmes and Toshi Yoshihara, "Getting Real about Taiwan," *The Diplomat*, March 7, 2011, <http://thediplomat.com/2011/03/07/getting-real-about-taiwan/?print=ye>.

[27] Bill. Owens, "American Must Start Treating China as a Friend," *Financial Times*, November 17, 2009, <http://www.ft.com/intl/cms/s/0/69241506-d3b2-11de-8caf-00144feabdc0.html#axzz2frtGJTnw>.

[28] Zbigniew. Brzezinski, "8 Geopolitically Endangered Species," *Foreign Policy*, January 3, 2012, <http://www.foreignpolicy.com/articles/2012/01/03/8_geopolitically_endangered_species?page=0,0>.

jectory toward Finlandization, which means Taiwan would reposition itself as a neutral power, rather than a U.S. strategic ally.[29] Charles Glaser makes similar arguments in analyzing Sino-American relations from international relations theory. The challenge for the U.S. will come in making adjustments to its policies in situations in which less-than-vital interests such as Taiwan might cause problems and in making sure it does not exaggerate the risks posed by China's growing power and military capabilities (Glaser 2011: 80-91).[30]

Denny Roy argues, on the contrary, abandoning Taiwan is completely at odds with the policy of Asia pivot or rebalancing (Roy 2012). Shelley Rigger further elaborates, phasing out U.S. security assistance to Taiwan would not only undermine vital security architecture in the Asia-Pacific region but also send a chilling message to our democratic allies around the world that American friendship is fickle (Rigger 2011). Michael Mazza emphasizes that Taiwan isn't a relic of the Cold War. Rather, it is situated at the geographic forefront of the strategic competition that very well may define the 21st century—that between the U.S. and China (Mazza 2011). John Copper observes, Asia pivot means the U.S. needs to check China's rise. Since the U.S. does not have money to finance a meaningful pivot, Washington has to seek allies. Washington granted Taiwan passport holders the right to enter the U.S. and remain for ninety days without a visa and supported Taiwan's participation in the International Civil Aviation Organization (ICAO). The U.S. moves clearly showed signs of a desire for better relations with Taiwan.[31]

---

[29] Bruce. Gilley, "Not So Dire Straits: How the Finlandization of Taiwan Benefits U.S. Security," *Foreign Affairs*, Vol. 89, No. 1 (2010), pp. 44-60.

[30] Charles. Glaser, "Will China's Rise Lead to War?" *Foreign Affairs*, Vol. 90, No. 2 (2011), pp. 80-91.

[31] Copper, John F., "Obama Turns toward Taiwan," *National Interest*, September 12, 2013, <http://nationalinterest.org/print/commentary/obama-turns-toward-taiwan-9048>.

Taiwan is even envisioned in China's geography as having geostrategic salience. The commonly accepted vision of Taiwan's strategic relationship to China entails the characterization of Taiwan as a "protective screen" (屏障, *pingzhang*) for China's southeastern coast and the Taiwan Strait as the crux for defense of the entire coastal region. Taiwan is described as the "strategic gateway of the southeast" (東南的鎖鑰, *dongnan de suoyue*). Taiwan and Hainan together constitute a "pair of eyes" (雙目, *shuangmu*) and, with the Zhoushan islands off the northeast coast of Zhejiang, constitute a strategic "horn" (犄角, *jijiao*) with Taiwan at the center. Taken together, these territories form naturally an advantageous battle array in the shape of the Chinese character *pin* (品) offering coastal defense sufficient to shield China's six southeastern coastal provinces and cities.[32]

Taiwan is seen as controlling a region denoted as the strategic "throat" (咽喉, *yanhou*) of the Western Pacific island chains, as the strongpoint closest to the mainland, and the demarcation of the boundary between the East China Sea (東海, *Donghai*) and South China Sea (南海, *Nanhai*). As such, Taiwan is a defensive outpost protecting the Chinese state, safeguarding China's north-south maritime traffic and the communications hub for China's maritime security region, and defending China's maritime rights and interest and blue territorial base. Once unification is achieved the PLA can again make use of Taiwan's territory, waters, and airspace for military bases in advance of activities in the Pacific oceanic and aerial battlefronts. So, Taiwan is both an asset in the hands of China and a liability in the hands of an adversary. It is described as having the potential to be both a bridge and springboard (橋梁和跳板, *qiaoliang he tiaoban*).[33]

---

[32] Alan M. Wachman, *Why Taiwan: Geostrategic Rationale for China's Territorial Integrity* (Stanford California: Stanford University Press, 2007), pp. 142-143.

[33] Wachman, *Why Taiwan: Geostrategic Rationale for China's Territorial Integrity*, p. 143.

Beyond protection of China's territory, Taiwan is thought to be of value in protecting vital sea lanes. Once unification is achieved, China's coastal defense can be pushed out eastward 300-500 kilometers, increasing the strategic depth at sea. This would enable the PLAN to widen its scope of operations to encompass the entire China Sea area, to access the East China Sea and the Yellow Sea from the south, to defend against adversaries entering China's northern maritime area from the north, and to strike back directly at any enemy penetrating the south, and access the Philippine Sea area directly from the west to deal with any enemy. By controlling Taiwan and the Taiwan Strait, Beijing could also thwart the efforts of enemies seeking to blockade China. The PLAN and PLAAF (PLA Air Force) could sweep through the length and breath of the Western Pacific maritime area, cut the American forward strategic chain in the Pacific (前沿戰略鏈環, *qianyan zhanlue lianhuan*), look down and control East Asia (瞰制東亞, *kanzhi dongya*), and control Japan's maritime lifeline to Southeast Asia. Moreover, the Diaoyu (Tiaoyutai, Senkaku) Islands to the north and the waters surrounding it would fall within the range of the PLA's gunfire. [34] It is for reasons of geographic realpolitik that China is determined to incorporate Taiwan into its dominion.

In brief, the island chains in the western Pacific were identified in 1949-1950 as the recently gained "forward defense perimeter" to be maintained for the future. The geopolitical problem for the U.S. is that retention of its forward defense perimeter now comes up against an emerging Chinese drive to achieve maritime penetration of the same perimeter line. [35]

---

[34] Wachman, *Why Taiwan: Geostrategic Rationale for China's Territorial Integrity*, pp. 143, 146.

[35] David Scott, "US Strategy in the Pacific—Geopolitical Positioning for the Twenty-First Century," *Geopolitics*, Vol. 17, No. 3 (2012), p. 617.

III. A2AD and ASCEL

The term of A2AD is a Western construct, and its approximation in the Chinese strategic concept is China's active strategic counterattacks on exterior lines (ASCEL) (積極的戰略外線反擊作戰, *jiji de zhanlue waixian fanji zuozhan*). Both concepts are similar and share many commonalities.

China started to understand the A2AD doctrine in the mid-1980s. Chinese planners began to shift away from planning for a war with the Soviet Union and began gradually to think about way to modernize the PLA and incorporate new technology and fighting doctrine. The 1991 Persian Gulf War sent shockwaves throughout China's military community and accelerated the PLA's modernization and shifts in strategy. In 1993, then-President Jiang Zemin ordered Chinese military planners to focus on preparing to wage "local wars under high technology conditions.".[36] This has been updated in 2004 to "local wars under condition of informationalization."[37] This would include two components: limited in geographical scope, duration, and political objectives and dominated by high technology weaponry. There are obviously other events, recent or otherwise, that Chinese planners looked at when crafting A2AD (which the Chinese referred to as counter-intervention operations). The wars in Bosnia, Kosovo and the 1995-1996 Taiwan Strait crises, as well as the 2001 Hainan Island incident, were all major factors for China when considering the development of its military strategy. The Taiwan Strait crisis especially holds significant weight, as China at the time had very little in the way of strategic options in countering an

---

[36] U.S. Department of Defense, *Military Power of the People's Republic of China 2000*, p. 5.

[37] Kathleen T. Rhem, "China Investing in Information Warfare Technology, Doctrine," *American Forces Press Service*, July 20, 2005,
<http://www.defense.gov/News/NewsArticle.aspx?ID=16594>.

American carrier off its coast, a reality that must have been a real spur to develop the DF-21D carrier kill missile.[38]

The 1997 *Quadrennial Defense Review* (QDR) tangentially expressed the concerns of A2AD, and stated "that if an adversary ultimately faces a conventional war with the U.S., it could also employ asymmetric means to delay or deny U.S. access to critical facilities; disrupt our command, control communication, and intelligence network"[39] The 2001 QDR identified missiles (both ballistic and cruise) and chemical, biological, radiological, nuclear and explosive (CBRNE) weapons as the greatest anti-access threats, particularly for their ability to deny or delay U.S. military access to overseas bases, airfields, and ports. Other anti-access threats of concern included advanced air defense systems that could threaten nonstealthy aircraft, and advanced mines, anti-ship cruise missiles (ASCM), and diesel submarines that could threaten the ability of U.S. naval and amphibious forces to operate in littoral waters.[40] The 2006 QDR specifically pointed out that "China has the greatest potential to compete militarily with the U.S. and field disruptive military technologies that could over time offset traditional U.S. military advantages absent U.S. counter strategies."[41] According to 2010 QDR, China is developing and fielding large numbers of advanced medium-range ballistic and cruise missiles, new attack submarines equipped with advanced weapons, increasingly capable long-range air defense systems, electronic warfare and computer network attack capabilities, advanced fighter aircraft, and counter-space systems. [42]

---

[38] Harry Kazianis, "An Anti-Access History Lesson," *The Diplomat*, May 25, 2012, <http://thediplomat.com/flashpoints-blog/2012/05/25/an-anti-access-history-lesson/?print=yes>.

[39] U.S. Department of Defense, *Quadrennial Defense Review Report*, 1997, p. 4.

[40] U.S. Department of Defense, *Quadrennial Defense Review Report*, 2001, pp. 26, 31, 42-43.

[41] U.S. Department of Defense, *Quadrennial Defense Review Report*, 2006, p. 29.

[42] U.S. Department of Defense, *Quadrennial Defense Review Report*, 2010, p. 31.

Moreover, in *the Military Power of the People's Republic of China 2004*, it first mentioned the concept of anti-access strategy, and denial and deception (D&D). Beijing sees Washington as principal hurdle to any attempt to use military force to regain Taiwan. China could consider a sea-denial strategy to hold at risk U.S. naval forces approaching the Taiwan Strait. Chinese D&D practices appear to be intended to delay or reduce U.S. diplomatic and military roles in crises.[43] Since then, the annual report has monitored China's A2AD developments on year-by-year basis. The Chinese A2AD military strategy includes the following key characteristics:

1. It aims to prevent friendly forces entry into a theater of operations and their freedom of action;[44]

2. It is intended to slow deployment of friendly forces and impede friendly operations;[45]

3. It is contemplated to conduct preemptive attacks designed to inflict severe damage on friendly forces based or operating in the western Pacific theater of operations;[46]

4. It is defensive Air-Sea Battle against offensive Air-Sea Battle;[47]

5. It is denial of sea control against assertion of sea control, [48]and equivalent to guerrilla warfare at sea;

6. It is an asymmetric military strategy.[49]

---

[43] Office of the Secretary of Defense, *Military Power of the People's Republic of China 2004*, p. 51.

[44] Krepinevich et al., *Meeting the Anti-Access and Area-Denial Challenge* (Washington D.C.: Center for Strategic and Budgetary Assessment, 2003), p. 5.

[45] Air-Sea Battle Office, *Air-Sea Battle: Service Collaboration to Address Anti-Access & Area Denial Challenges*, 2013, p. 2.

[46] Tol et al., *Airsea Battle: A Point-of-Departure Operational Concept* (Washington D. C.: Center for Strategic and Budgetary Assessment, 2010), p. xii.

[47] Robert C. Rubel, "Talking About Sea Control," *Naval War College Review*, Vol. 63, No. 4 (2010), p. 40.

[48] Stansfield. Turner, "Missions of the U.S. Navy," *Naval War College Review*, Vol. 26, No. 5 (1974), pp. 2-17.

Whereas the term A2AD is a Western construct, its approximation in the Chinese strategic concept is the PLA's ASCEL. In the U.S. perspective, Beijing's view of active offshore defense includes the more strategic label of ASCEL. The important factor in any Chinese formulation is that defense does not mean defense in U.S. terms. Rather, active defense is best understood as "we will attack when we see an advantage in doing so".[50] The core nature of ASCEL operations is encapsulated by the following three points:

1. It is not an operation on the exterior lines at the level of campaign or combat, but an operation conducted at the level of strategy;

2. It is not an operation in the phase of strategic counterattack within the three phases of general war (strategic defense, strategic stand-off, strategic counterattack), but is a strategic operation conducted from the very beginning;

3. It is a strategically defensive and active self-defense counterattack and a component of the strategy of active defense.[51]

China's ASCEL operations and A2AD share numerous commonalities, including their strategic level as well as their integration with overall national military strategy. In addition to these rather general similarities, there are number specific points of commonality. For one, both place special emphasis on striking the enemy in the early stages of a conflict as well as preventing the enemy's approach, development, and power projection within a specific theater. Though China's ASCEL operations are classified as a counterattack, they are carried out from the very beginning of a conflict, emphasize early, active offensive operations conducted as far as possible from China's terri-

---

[49] Robert C. Manke, and Raymond J. Christian, *Asymmetry in Maritime Access and Undersea Anti-Access/Area-Denial Strategies* (New Port, Rhode Island: Naval Undersea Warfare Center, 2007).

[50] Bernard D. Cole, "Controlling Contested Water," *Proceedings Magazine*, Vol. 139, No. 10 (2013), p. 52.

[51] Anton Lee Wishik II, "An Anti-Access Approximation," *China Issue*, Issue 19 (2011), p. 40.

tory, and can potentially consist of the first shot at the tactical level. In addition, both concepts are decidedly asymmetric in nature and emphasize the defeat of a militarily superior enemy by a weaker side. Furthermore, they both place great emphasis on the naval role. Aerial and long-range strike forces due to the unlikelihood of land invasion and the relevant geographical scope and military targets, namely attacks on overseas military bases, battle platforms and deployment systems. It is also true that both share a common trigger for their activation, namely the perceived harming of China's core interests, territorial integrity and sovereignty. Finally, perhaps most compelling is the fact that they both share a similar geographical scope, namely the area between and immediately surrounding the first and second island chains.[52]

IV. China's A2AD Threats

The PLA's ASCEL operations and A2AD provide a doctrinal-based starting point for an A2AD assessment, but military capabilities constitute a greater proportion of evidence in support of such as assessment than actual PLA strategic writings, and fill the gaps between rhetoric and reality.

In July-August 1995 and March 1996, concerns about Taiwanese President Lee Teng-hui's measures that Chinese leaders associated with moves toward de jure independence of Taiwan led Beijing to conduct missile tests and other military exercises near the Strait. To deter further escalation, then U.S. President William Clinton dispatched two carrier strikes groups (CSGs) toward the region in March 1996, later remarking, "When word of crisis breaks out in Washington, it is no accident the first question that comes to everybody's lips is: where is the nearest carrier?"[53] In the unfortunate event of a future Sino-U.S. military crisis, however, it is Chinese leaders who

---

[52] Wishik II, "An Anti-Access Approximation," p. 44.

[53] "The Carriers—Why the Carriers," *America's Navy*, <http://www.navy.mil/navydata/ships/carriers/cv-why.asp>.

would be asking where the nearest U.S. carrier is, albeit for the opposite reason.[54]

Since 1996, China has developed and acquired the technologies that could hold U.S. and allied military platforms and their supporting assets at risk in the western Pacific, and sought to deter an American carrier, a potent symbol of U.S. military might, from plying the edge of Chinese waters once again. In July 2010, Beijing opposed joint U.S.-South Korean military exercises in the Yellow Sea, which were participated in by the carrier USS *George Washington* in response to the March sinking of the South Korean patrol ship *Cheonan*. Beijing protested so vociferously that the U.S. and South Korea shifted planned maneuvers to the Sea of Japan, east of South Korea.[55]

At present, China's A2AD strategy is ever more potent. According to the *Military and Security Developments Involving the People's Republic China 2013*, these A2AD capabilities have been dramatically enhanced in the following trends: Firstly, cyberwarfare attacks and other espionage efforts. China has long been accused of engaging in cyber attacks and espionage on US networks, but for the first time the US military directly attributed some of those attacks to the PLA. Secondly, use of space to thwart the US military. In 2012, China conducted 18 space launches to expand its intelligence and surveillance satellites. China is working quickly to improve its capabilities to limit or prevent the use of space-based assets by adversaries during times of crisis or conflict. Thirdly, development of carrier-killer missiles. China is developing specialized, precision anti-ship ballistic missiles

---

[54] Andrew S. Erickson, "China's Evolving Anti-Access Approach: Where's the Nearest (U.S.) Carrier?" *China Brief*, Vol. 10, No. 18 (2010), p. 5.

[55] Jeremy Page et al., "China Warns U.S. as Korean Tensions Rise," *Wall Street Journal*, November 26, 2010,
<http://online.wsj.com/article/SB10001424052748704008704575638420698918004.html>
.

(ASBM) that are capable of hitting US aircraft carriers from a range exceeding 1,500 kilometers. Finally, development of sophisticated ships, planes, and drones. China is developing fourth- and fifth-generation aircraft that incorporate stealth technology. China has also launched its first aircraft carrier. The formation of carrier battle groups will enable the PLAN to conduct comprehensive operations and enhance its long-range operational capabilities. The PLA is investing heavily in a robust program for undersea warfare, developing nuclear-powered attack submarines. China is also interested in expanding its fleet of drones.[56]

Cyber attacks are new dimension of the ongoing strategic competition between the U.S. and China, although cyberwarfare is nothing new. The PLA revised its doctrine from local wars under high technology conditions to local wars under informationalized conditions in 2004, while the U.S. National Security Agency (NSA) received the mission for Computer Network Attack (CNA)—offensive cyberwarfare—on March 3, 1997, from then Defense Secretary William S. Cohen. The future of warfare is warfare in cyberspace. The primary target of this option is the information infrastructure of an adversary. Such information infrastructures are expected to be primarily computer controlled, operated by the commercial-civilian sector (unprotected), and the primary infrastructure upon which military forces almost totally depend. As a result, information warriors will need to be expert in understanding the virtual world and have extensive knowledge of non-military targets. Military cyberwarriors will be the tooth, and civilians will be the tail in what the military calls the tooth-to-tail—frontline and support—relationship in warfare.[57]

---

[56] Office of the Secretary of Defense, *Military and Security Developments Involving the People's Republic of China 2013*, p. i.

[57] Bill. Gertz, "Inside the Ring: NSA on Cyberwar," *Washington Time*, March 27, 2013, <http://www.washingtontime.com/news/2013/mar/27/inside-the-ring-nsa-on-cyberwar>.

Consider what might happen in a broader U.S.-China conflict. The PLA could conduct major efforts to disable critical U.S. military information systems. Even more ominously, PLA cyberwarriors could turn their attention to strategic attacks on critical infrastructure in America.[58] However, "We believe our cyber offense is the best in the world", said General Keith B. Alexander, Director of the NSA and Commander of U.S. Cyber Command.[59] China's military fears a major cyberattack against its strategic forces and communist leaders also worry about cyberstrikes against infrastructure. A devastating cyberattack on its military or civilian infrastructure is one of Beijing's 16 strategic fears.[60]

Of perhaps greatest concern, by December 2012, Beijing had a formidable arsenal of 1,100 short-range ballistic missiles (SRBM) deployed to units opposite Taiwan,[61] while in 2002 it had only 350.[62] China is also fielding a limited but growing number of conventionally armed, medium-range ballistic missiles (MRBM). As China's ability to deliver accurate fire across the strait grows, it is becoming increasingly difficult and soon may be impossible for Taiwan and the U.S. to protect the island's military and civilian infrastructures from serious damage. China's ability to suppress Taiwan and local U.S. air bases with ballistic and cruise missiles seriously

---

[58] Dan. Blumenthal, "How to Win a Cyberwar with China," *Foreign Policy*, February 28, 2013,
<http://www.foreignpolicy.com/articles/2013/02/28/how_to_win_a_cyberwar_with_china?print=yes&hidecomments=yes&page=full>.

[59] Steven. Aftergood, "US Cyber Offense Is the Best in the World," *Secrecy News*, August 26, 2013, <http://blogs.fas.org/secrecy/2013/08/cyber-offense>.

[60] Michael. Pillsbury, "The Sixteen Fears: China's Strategic Psychology," *Survival*, Vol. 54, No. 5 (2012), pp. 158-159.

[61] Office of the Secretary of Defense, *Military and Security Developments Involving the People's Republic of China 2013*, p. 5.

[62] Office of the Secretary of Defense, *Military Power of the People's Republic of China 2002*, p. 2.

threatens the defense ability to maintain control of the air over the strait. Worst of all, the U.S. can no longer be confident of winning the battle for the air in the air. This represents a dramatic change from the first six-plus decades of the Taiwan-China confrontation.[63]

An unclassified Defense Intelligence Agency report assessing the state of Taiwan's air defenses raises similar concerns. The report notes that despite the operational capability of Taiwan's fighter force, these aircraft cannot be used effectively in conflict without adequate airfield protection, especially runways, suggesting a major vulnerability to the island's airpower. Taiwan's ability to protect its aircraft and airfields from missile attacks and rapidly repair damaged runways and taxiways are central issues to consider when examining Taiwan's air defense capability.[64]

China's missiles also threaten Taiwan's ability to defend itself at sea. William Murray contends that China could sink or severely damage many of Taiwan's warships docked at naval piers with salvos of ballistic missiles. He argues that the Second Artillery's expanding inventory of increasingly accurate SRBM would probably allow Beijing to incapacitate much of Taiwan's navy and to ground or destroy large portions of the air force in a surprise assault and follow-on barrages.[65]

Equally troubling is growing evidence that China has turned its attention to Japan, home to some of the largest naval and air bases in the world, e.g., Yokosuka, Sasebo, Kure, Maizuru, Kadena, and Misawa. Beijing has

---

[63] David A. Shlapak et al., *A Question of Balance: Political Context and Military Aspects of the China-Taiwan Dispute* (Santa Monica California: Rand, 2009), pp. 126, 131, 139.

[64] Defense Intelligence Agency, *Taiwan Air Defense Assessment*, DIA-02-1001-028, January 21, 2010, <http://www.globalsecurity.org/military/library/report/2010/taiwan-air-defense_dia_100121.htm>.

[65] William S. Murray, "Revisiting Taiwan's Defense Strategy," *Naval War College Review*, Vol. 61, No. 3 (2008), p. 24.

long worried about Tokyo's potential role in a cross-strait confrontation. In particular, Chinese analysts chafe at the apparent American freedom to use the Japanese archipelago as a springboard to intervene in a Taiwan contingency. In the past, China kept silent on what the PLA would do in response to Japanese logistical support of U.S. military operations. Recent PLA publications, in contrast, suggest that the logic of missile coercion against Taiwan could be readily applied to the U.S. forward presence in Japan.[66]

Evidence suggests that China's emerging strategy is actually much more ambitious, direct and therefore dangerous for the U.S. One of characteristics for the A2AD strategy is the wide proliferation of long-range ballistic and cruise missile technologies and the convergence of Chinese military power around a missile-centric, rather than the conventional platform-centric, model of mass-firepower combat. Missiles are cheap, fast, expendable, risk no friendly casualties and, most importantly, are difficult to preempt. Moreover, they do not require air superiority to operate and offer a high, often uninhibited, rate of defense penetration. China can thus use missiles not only to achieve strategic surprise but to dismember U.S. assets on the ground or at sea without putting its own hardware or personnel in harm's way. For this reason, missiles have permeated the PLA's doctrine for every important kind of operation, from denial to blockade, and the PLA officer corps views them more and more as the way to level the playing field against a superior adversary.[67]

In addition to attacking land-based targets, another weapon that warrants discussion is, of course, DF-21D ASBM. Given China's overall inferiority in long-range air and naval power, an ASBM would afford a powerful asymmetric means that could deter the U.S. forces on their way to a zone of

---

[66] Yoshihara, "Chinese Missile Strategy and the U.S. Naval Presence in Japan," p. 40.

[67] Vitaliy O. Pradun, "From Bottle Rockets to Lightning Bolts: China's Missile Revolution and PLA Strategy against U.S. Military Intervention," *Naval War College Review*, Vol. 64, No. 2 (2011), p. 11.

conflict near China's littoral borders. However, the ASBM represents more than just a single weapon platform. Rather, it is seen as "a system of systems" and a key step in achieving high-tech and information war capabilities. This is because the ability to launch a land-based ballistic missile at a moving target thousands of kilometers away requires a wide range of support and information technologies far beyond just the missile itself. Certainly, the ASBM is the core component of this system, and the technological demands in maneuvering, guidance, and homing to defeat defenses and find its moving target at sea are formidable. Nonetheless, an effective ASBM would also require the ability to detect, identify, and track the target using some combination of land, sea, air, and space-based surveillance assets. Aside from the immediate software and hardware, all of these functions would have to be highly integrated, fast reacting, and sufficiently flexible to attack the world's most sophisticated and best defended naval target in the world today—an American CSG.[68]

Moreover, the DF-21D would take approximately thirty-five minutes from the detection of the target for the PLA to communicate its location to a relevant C2 center, issue an engagement order (with no delay assumed) to the launcher, and fire the ASBM, and for the missile to travel its full range. During the thirty-five minutes the carrier group could travel thirty-one kilometers, making a circle with a radius of thirty-one kilometers the missile's area of uncertainty and therefore the required seeker footprint for a single missile to find the target.[69] Although no authoritative data on the DF-21D's seeker footprint exist in the open literature, Chinese sources suggest twenty-,

---

[68] Eric Hagt and Matthew Durnin, "China's Antiship Ballistic Missile: Developments and Missing Links," *Naval War College Review*, Vol. 62, No. 4 (2009), p. 87.

[69] Marshall Hoyler, "China's Antiaccess Ballistic Missiles and U.S. Active Defense," *Naval War College Review*, Vol. 63, No. 4 (2010), pp. 93-94.

forty-, and hundred-kilometer footprints.[70] Given the missile's high cost, it is unlikely that China would opt for an overly narrow footprint, making a hundred, or perhaps forty, kilometers more credible than twenty. Hence, chances are that each individual ASBM would be able to find its target and, once it does, achieve a virtually assured hit. [71]

The U.S. Navy conceded in December 2010 that DF-21D had reached initial operating capability. Admiral Jonathan Greenert, the Navy's top officer, reveals that the sailors are working out several different options to kill it before it kills them. Some involve convincing the DF-21D that the carrier is in a different place. Others involve masking the electronic emissions of the carrier. Still others are more traditional—like blasting the missile out of the salty air. "You want to spoof them, preclude detection, jam them, shoot them down if possible, get them to termination, confuse it," Greenert said.[72]

Although China's anti access—Defensive Air-Sea Battle—capability is impressive, significant holes remain in the PLAAF modernization. Foremost among these is its small air refueling fleet. China has perhaps eight Il-78 tankers and may have converted up to a dozen H-6 bombers to refueling status.[73] The PLAN continues to exhibit weakness in several areas, including capabilities for sustaining operations by larger formations in distant waters, joint operations with other part of the PLA, antisubmarine warfare (ASW),

---

[70] Hoyler, "China's Antiaccess Ballistic Missiles and U.S. Active Defense," p. 93; Eric Hagt and Matthew Durnin, "China's Antiship Ballistic Missile: Developments and Missing Links," pp. 87-115.

[71] Pradun, "From Bottle Rockets to Lightning Bolts: China's Missile Revolution and PLA Strategy against U.S. Military Intervention," p. 25.

[72] Spencer Ackerman, "How to Kill China's Carrier-Killer Missile: Jam, Spoof and Shoot," *Wired Danger Room*, March 16, 2012, <http://www.wired.com/dangerroom/2012/03/killing-chinas-carrier-killer>.

[73] Rebecca. Grant, "Meet the New PLAAF," *Air Force Magazine*, Vol. 96, No. 1 (2013), p. 37.

mine countermeasures (MCM), etc.[74] Both PLAAF and PLAN have a dependence on foreign suppliers for their propulsion systems and a lack of combat experience.

V. Breakthrough First Island Chain

China's growing interest in the Ryukyu island chain, in particular the southern Sakishimas, has paralleled its own growing capabilities and ambitions. Prior to 2008, there were very few PLAN activities in the Ryukyu region. However, since 2008, particularly in 2013, they have become a regular occurrence.

The year of 2008 is a year of the PLAN shifted from the near-seas navy to the far-seas navy. China builds up its surface warships rapidly and is eager to flex its maritime military muscle (Table 2). In 2005 the PLAN only commissioned 9 modern destroyers and 1 frigate.[75] In 2008 the PLAN commissioned 13 destroyers, 3 frigates and 1 amphibious transport dock (LPD),[76] while in 2013 it commissioned 15 destroyers, 14 frigates and 3 LPDs.[77] In December 2008, in addition to sending warships eastward to break through the first island chains, China dispatched a three-ship flotilla southward to the Gulf of Aden to protect merchant ships from Somali pirate attacks. Since then, the PLAN has rotated its counter-piracy escort flotilla and frequented the Ryukyu region. In comparison, counter-piracy escort is military operations other than war (MOOTW), while break-through the first island chains has more geostrategic implication in the Washington-Tokyo vs. Beijing competition.

---

[74] Ronald. O'Rourke, *China Naval Modernization: Implications for U.S. Navy Capabilities*, RL33153 (Washington D.C.: Congressional Research Service, 2013), p. 3-4.

[75] Ships of the World, *World's Navy 2005-2006*, 2005, pp. 35-38.

[76] Ships of the World, *World's Navy 2008-2009*, 2008, pp. 30-34.

[77] Ships of the World, *World's Navy 2013-2014*, 2013, pp. 31-34.

Chinese strategists and scholars have a hot debate on which strategic direction should go first when the PLAN projects its power. Fudan University's Shen Dingli explains, "For the East China Sea, it is more political. China considers we have been invaded by Japan and Japan has stolen our Diaoyu Islands. But for the South China Sea, it is largely about economics".[78] Retired Rear Admiral Yangyi opines, "Facing the enemy's encirclement, economics may act as a vanguard and military power as a rearguard. 'March South' has better served China's long term interests and will face weak neighboring states".[79] Yang further elaborates that China's expansion into the Pacific and Indian Oceans is a prerequisite for the country to call itself a great global power.[80]

China's naval presence in the Gulf of Aden has recalled the historical sea trade route that extended coast-wise through the China Sea and the Southeast Asian archipelago to India, Arabia, Africa, and perhaps even Australia some 300 years before the arrival of Captain Cooke. In the 15th century, the Chinese navigator Zheng He made seven trade voyages to the "Western Seas" and established a moment in history when China ruled the seas as both economic and naval power.[81] In October 2013, President Xi Jinping was in Malaysia and Indonesia. During his trip in Indonesia, he said in a speech that China and the ASEAN (Association of Southeast Asian Nations)

---

[78] Zachary. Keck, "Why Is China Isolating Japan and the Philippines?" *The Diplomat*, October 26, 2013,
<http://thediplomat.com/flashpoints-blog/2013/10/26/why-is-china-isolating-japan-and-the-philippines>.

[79] "Hong Kong Media: China's Grand Strategy Faces Differences (港媒：中國大戰略面臨分歧)," *Ta Kung Pao*, November 6, 2012,
<http://news.takungpao.com/military/view/2012-11/1254237.html>.

[80] Yun. Sun," Westward Ho," *Foreign Policy*, February 7, 2013,
<http://www.foreignpolicy.com/articles/2013/02/07/westward_ho_china_asia_pivot#sthash.G6fAEe7F.8iDGkv4D.dpbs>.

[81] Peter. Neill, "Maritime Silk Road," *Huffington Post*, October 31, 2013,
<http://www.huffingtonpost.com/peter-neill/maritime-silk-road_b_4181663.html>.

will promote maritime cooperation and build a 21st-century new "maritime Silk Road" (MSR). China tries to promote the MSR to crack the possible American blockades.[82]

However, earlier MSR was used for the import of precious stone, wood and spices but today it will be used for oil and gas, which is directly connected to the energy security of not one but many countries. There is emerging security architecture in the region which has led to an increased arms buildup, and the assertiveness of new regional powers has complicated the regional military balance, which makes a revival of the MSR looks bleak.[83]

Table 2: Chinese Warship Activities in the Vicinity of Japanese Islands

Date	Warship Class	Activity
Oct. 2008	A *Sovremenny*-class destroyer and four other vessels	Passed through the Miyako Strait from the Pacific Ocean after transited the Tsugaru Strait.
2 Nov. 2008	Four surface vessels, including a *Luzhou*-class destroyer	Passed through the Miyako Strait on their way to the Pacific Ocean
25 June 2009	A *Luzhou*-class destroyer and four other vessels	Traversed the Miyako Strait
18 March 2010	Six warships, including a *Luzhou*-class destroyer	Passed through the Miyako Strait to the Pacific Ocean
22 April 2010	Ten warships including two *Sovremenny*-class destroyers and two *Kilo*-class submarines	Passed through the Miyako Strait, during which time a Chinese helicopter buzzed a Japanese destroyer

---

[82] Simon. Denyer, "China Bypasses American 'New Silk Road' with Two of Its Own," *Washington Post*, October 14, 2013, <http://www.washingtonpost.com/world/asia_pacific/china-bypasses-american-new-silk-road-with-two-of-its-own/2013/10/14/49f9f60c-3284-11e3-ad00-ec4c6b31cbed_print.html>.

[83] Teshu. Singh, "China and ASEAN: Revisiting the Maritime Silk-Road—Analysis," *Eurasia Review*, October 15, 2013, <http://www.eurasiareview.com/15102013-china-asean-revisiting-maritime-silk-road-analysis>.

3 July 2010	Two vessels, including a *Luzhou*-class destroyer	Passed through the Miyako Strait
8-9 June 2011	Eleven vessels, including three *Sovremenny*-class destroyers	Transited through the Miyako Strait to the Pacific Ocean. On 22 June, the same eleven vessels transited back to the East China Sea
22 Nov. 2011	Six vessels including a *Luzhou*-class destroyer and a *Luhu*-class destroyer	Passed through the Miyako Strait to the Pacific Ocean. On Dec. 1, five vessels transited back to the East China Sea
2 Feb. 2012	A *Jiangkai II*-class and three *Jiangwei*-class frigates	Passed through the Miyako Strait to the Pacific Ocean. On Feb. 10, the same four vessels transited back to the East China Sea
6 May 2012	Five vessels including two *Luyang I*-class destroyers and two *Jiangkai II*-class frigates	Sailed southeastward in international waters about 650 km southwest of Okinawa
15 May 2012	Three vessels including two *Jiangkai II*-class frigates and one intelligence collection ship	Passed through the Miyako Strait to the East China Sea. On 29 April, the same vessels transited the Osumi Strait
23 June 2012	Three vessels including a *Luzhou*-class destroyer and a *Jiangwei II*-class frigate	Passed through the Miyako Strait to the East China Sea. On 13 June, the same vessels transited the Osumi Strait
4 Oct. 2012	Seven vessels including two destroyers, two frigates, two submarine rescue ships and one supply ship	Transited through the Miyako Strait to the Pacific Ocean. On 16 October, the same seven vessels transited back to the East China Sea through the Taiwan-Yonaguni Strait
23 Oct. 2012	Three vessels including two destroyers and a frigate	Transited through the Miyako Strait from the Pacific Ocean. These ships entered the Pacific Ocean from southern PLAN bases
28 Nov. 2012	Four vessels including two guided-missile destroyers, and two missile frigates	Passed through the Miyako Strait on their way to the Pacific Ocean. On December 10, the same ships transited back to the East China Sea through the Taiwan-Yonaguni Strait

31 Jan. 2013	Three vessels including a *Luhu*-class destroyer and two *Jiangkai II*-class frigates	Transited through the Miyako Strait to the Pacific Ocean. On 13 February, the same vessels transited back to the East China Sea
2 April 2013	A *Luyang II*-class destroyer and two *Jiangkai II*-class frigates	Sailed westward in international waters about southwest 650 km of Okinawa. On 31 March, the same vessels were spotted sailing eastward in the same waters
16 April 2013	A *Luyang II*-class destroyer and a *Jiangkai II*-class frigate	Passed through the Miyako Strait on their way to the Diaoyu Islands waters
13 May 2013	A *Jiangwei II*-class frigate and a *Jianghu V*-class frigate	Headed westward in international waters about 660 km southwest of Okinawa. On 7 May, the same ships were spotted about 44 km northeast of Yonaguni
27 May 2013	A *Luhu*-class destroyer, a *Liangkai II*-class frigate and a supply ship	Passed through the Miyako Strait and entered the Western Pacific for a training mission
8 June 2013	A *Luhu*-class destroyer and a *Jiangwei II*-class frigate	Transited through the Osumi Strait. In early June, the same vessels were spotted in international waters about 450 km south of Raso Island
25 July 2013	Two *Luzhou*-class destroyers, two *Liangkai II*-class frigates, and a supply ship	Passed through the Miyako Strait and entered the East China Sea. These ships have accomplished the Sino-Russian joint exercise and on July 14, transited the Soya Strait on their way to the Pacific Ocean to conduct training exercises
21 Aug. 2013	Three vessels including a *Luhu*-class destroyer, a *Jiangkai II*-class frigate and a supply ship	Transited through the Osumi Strait on their way to Hawaii. On October 30, the same vessels transited through the Miyako Strait back to the East China Sea
27 Aug. 2013	Two *Jiangkai II*-class frigates	Passed through the Miyako Strait to the Pacific Ocean. On 9 September, the same vessels transited back to the East China Sea

23 Oct. 2013	Five vessels including two *Luzhou*-class destroyers and three *Liangkai II*-class frigates	Passed through the Miyako Strait to the Pacific Ocean. On 29 October, three ships transited back through the Miyako Strait, and the other two ships were spotted about 44 km northeast of Yonaguni
30 Oct. 2013	A *Luyang I*-class destroyer, a *Jiangkai II*-class frigate, and a supply ship	Were spotted about 610 km southwest of Okinawa. On Oct. 23, the same ships passed through the Bashi Channel between Taiwan and the Philippines
23 Dec. 2013	Three vessels including two *Jiangkai II*-class frigates and a supply ship	Were spotted about 610 km southwest of Okinawa

Sources: Japan Ministry of Defense 2008-2013.

China's penetrating through the Miyako Strait is the symbol of national pride and patriotism. In addition to Chinese surface fleet activities, the Japan Maritime Self Defense Force (JMSDF) has tracked several Chinese submarine passages through Japanese straits in recent years (Table 3). Of great concern, on 9 November 2004, a Chinese *Han*-class nuclear-powered submarine entered Japanese territorial waters while submerged. The submarine moved through a Corridor between Ishigaki and Miyako islands at around 5:50 a.m., breaching Japanese territorial waters for about two hours. Since the 1990s, the PLAN has been exploring submarine routes that will take vessels to the Pacific between Taiwan and Okinawa.[84]

---

[84] "China Sub Tracked by U.S. off Guam before Japan Intrusion," *Japan Times*, November 17, 2004, <http://www.japantimes.com.jp/news/2004/11/17/national/china-s..>.

Table 3: Chinese Submarine Activities in the Vicinity of Japanese Islands

Date	Submarine Class	Activity
Nov. 2003	Confirmed: *Ming*-class diesel attack submarine	Passes through the Osumi Strait on the surface and flies the PRC flag
Nov. 2004	Confirmed: *Han*-class nuclear attack submarine	Passes submerged through Ishigaki Strait
Oct. 2006	Confirmed: *Song*-class diesel attack submarine	Surfaces within torpedo range of USS *Kitty Hawk* during exercises in the Pacific east of the Japan
Sept. 2008	Unconfirmed: PLAN submarine, unknown class	Passes submerged in Japanese territorial sea in the vicinity of Shikoku and Kyushu islands
Oct. 2008	Unconfirmed: *Han*- and *Song*-class attack submarines	Sit submerged in Japanese exclusive economic zone as USS *George Washington* passes en route to Busan
April 2010	Conformed: two *Kilo*-class Diesel attack submarines	Passes through the Miyako Strait with other vessels
May 2013	Confirmed: *Yuan*-class diesel attack submarine and unknown class	Passes submerged on May 2 in the west of Amami Oshima Island in Kagoshima prefecture; Passes submerged on May 12 in the south of Kume jima Island in Okinawa; Passes submerged on May 19 in the south of Minamidaito jima Island in Okinawa.

Sources: Peter Dutton, "Scouting, Signaling, and Gatekeeping: Chinese Naval Operations in Japanese Waters and the International Law Implications," *China Maritime Studies*, No. 2 (2009), p. 7; Matthew M. Burke and Hana Kusumoto, "Suspected Chinese Subs Raise Concerns in Japan," *Stars and Stripes*, May 23, 2013, <http://www.stripes.com/news/suspected-chinese-subs-raise-concerns-in-japan-1.222238>.

Peter Dutton argues that the appearance of Chinese submarines in the vicinity of Japanese island have the following strategic significance: Firstly, scouting: the action could have been a covert mapping exercise. For years the U.S. has been aware that the PLAN has been exploring various submarine routes through which to move its submarines into the central Pacific in the event of regional conflict. Some observers have suggested that relaxations in trade and technology restrictions in the 1990s allowed China to purchase advanced oceanographic mapping systems that enable it to make sophisticated maps of ocean floor. These maps could be very useful to the PLAN submarine force in the event of war. Additionally, the maps could be useful in exploring the seabed for suitable locations to drill and explore for gas and oil. Secondly, signaling: the PLAN has been demonstrating its sea power. It is conceivable that the submarine's submerged passage through Japanese waters was an intentional provocation to demonstrate to Japan and the U.S. the extent of the Chinese sea power and its blue-water capability, and possibly to test the military capabilities of the Japanese. To the U.S., China's message has consistently been that it should refrain from military support of Taiwan. To Japan, China's message may be related to ongoing maritime boundary and resource disputes. Finally, gatekeeping: access and denial of access during an East Asian crisis. In a military crisis over the status of Taiwan, one role for China's potent submarine force would be to support a PLA blockade of the island and to prevent the U.S. and Japan from using the choke points created by Japanese islands to deny Chinese vessels access to and from the Sea of Japan and the East China Sea during the period of crisis.[85]

What explains China's increased military activities near the Ryukyus? Beijing's 2012 defense white paper discloses that China's armed forces shall

---

[85] Peter Dutton, "Scouting, Signaling, and Gatekeeping: Chinese Naval Operations in Japanese Waters and the International Law Implications," *China Maritime Studies*, No. 2 (2009), pp. 19, 21-22.

unswervingly implement the military strategy of active defense, guard against and resist aggression, contain separatist forces, safeguard borders and coastal and territorial air security, and protect national maritime rights and interests and national security interests in outer space and cyber space.[86] China's ability to achieve these objectives is connected to the Ryukyus in various ways. First, Beijing's pledge to defend its land and territorial waters includes the disputed Diaoyu Islands, inevitably generating friction with Japan. Second, the maritime rights it desires to protect involve access to sea lanes, including the vital straits that connect the Ryukyus to the Pacific. Finally, the proximity of the Ryukyus to Taiwan means that should China resort to force to prevent Taiwanese independence, the Ryukyus would likely play a critical operational role. If China aimed to seize the Diaoyu Islands or other islands in the Sakishima chain during a Taiwan crisis, it would be forced to vie for sea control with Japanese and the U.S. forces.[87]

One key factor that affects security in the Ryukyu Islands is the PLA's amphibious capabilities. China seeks to project power ashore in Taiwan or various islands of the East China Sea. To date, its capabilities have been concentrated opposite Taiwan. The U.S. Department of Defense reported in 2013, however, that "the PLA is capable of accomplishing various amphibious operations short of a full-scale invasion of Taiwan. With few overt military preparations beyond routine training, China could launch an invasion of small Taiwan-held islands such as Pratas or Itu Aba. A PLA invasion of a medium-sized, better defended offshore island such as Matsu or Jinmen is within China's capabilities."[88] As a result, while China's amphibious assets

---

[86] PRC's Information Office of the State Council, *The Diversified Employment of China's Armed Forces*, 2013,
<http://eng.mod.gov.cn/Database/WhitePapers/2013-04/16/content_4442752.htm>.

[87] Eric Sayers, "The Consequent interest of Japan's Southwestern Islands," *Naval War College Review*, Vol. 66, No. 2 (2013), pp. 54-55.

[88] Office of the Secretary of Defense, Military and Security Developments Involving the People's Republic of China 2013, p. 57.

remain focused on the Taiwan Strait, they appear capable of assaults against small, lightly defended islands, potentially including Miyako, Ishigaki, or Yonaguni.[89]

China's maturing A2AD capability has threatened Taiwan, Japan and the U.S. forces stationed in the Western Pacific. One quick fix for the U.S. and its partner nations to counter the PLA's powerful military might is to bolster their version of A2AD capability. Taiwan's military has tried to meet China's threat asymmetrically. The tool for Taiwan's denial of access strategy includes the supersonic *Hsiung Feng III* (*Brave Wind III*) anti-ship missile (ASM). The missiles are designed to cruise at a speed of Mach 2 with a range of up to 130 kilometers.[90] The *Hsiung Feng III*-equipped patrol boats stationed in Keelung port will have combat radius easily within the Diaoyu Islands. The U.S. is even developing the Long Range Anti-Ship Missile (LRASM) for its naval and air forces. The LRASM provides the U.S. military with an offensive anti-surface weapon (OASuW) to strike the growing threats from A2AD strategy. Its 500 nautical miles range is crucial to restore the balance of hitting power. The LRASM will have air and surface-launched capability, travel at subsonic speed, and carry a 1,000-pound penetrator and blast-fragmentation warhead.[91]

Moreover, a Rand report suggests the U.S. military consider turning China's A2AD doctrines on its head by incorporating a far blockade strategy using land-based ASMs at chokepoints in the first island chain. The report argues that land-based ASMs would not only have a significant effect on China's ability to project power, but it would also vastly expand the set of military problems that the PLA would face should it consider launching a

---

[89] Eric Sayers, "The Consequent interest of Japan's Southwestern Islands," p. 56.

[90] Gavin. Phipps, "RoCN Corvettes Being Fitting with HF-3 Missiles," *Jane's Defense Weekly*, Vol. 49, No. 21 (2012), p. 16.

[91] Defense Advanced Research Projects Agency, *LRASM Prototype Scores 2nd Successful Flight Test*, 2013, <http://www.darpa.mil/NewsEvents/Releases/2013/12/03.aspx>.

conflict with its neighbors or U.S. allies. Land-based ASMs are so easy to operate and are strategically and tactically mobile. ASMs could be placed in many locations over thousands of miles of island chains, which would dilute the effectiveness of PLA missiles and air forces. If Taiwan and Japan became involved in a conflict with China, ASMs with an effective range of only 100-200 kilometers stationed on the island of Okinawa and northern Taiwan could cover all PLA naval traffic south of Okinawa. The Luzon Strait between the Philippines and Taiwan could be covered with 100 kilometers range ASMs positioned in Taiwan and the Philippines.[92] The spiral competition between access and anti access has intensified in the Western Pacific region since China has become more assertive and aggressive.

VI. Conclusion

While countervailing an emerging China, the essence of the U.S. Asia pivot or rebalancing is reviving again the geopolitics of the Western Pacific. China has a geostrategic preponderance on the continent. However, passage in and out of the open sea is blocked by two island chains. China's maritime geostrategic posture is thus in an easily encircling condition. Although the U.S. has the difficulty of projecting power overseas across extended lines of communication, its policy in the Pacific continues to maintain forward bases, position assets around chokepoints and the SLOCs, deploy a naval presence on all seas, and keep the capability to intervene at key flash points. Thus, in the maritime-continental competition, Taiwan is seen as an unsinkable carrier or a cork in China's bottle by the West, pinning Beijing's attention on the cross-strait stalemate. Taiwan is also seen as a protective screen or the strategic gateway of the southeastern provinces by the Chinese, offering coastal defense for their soft underbelly.

---

[92] Kelly et al., Employing Land-Based Anti-Ship Missiles in the Western Pacific, (Santa Monica, California: Rand, 2013).

Whereas the term of A2AD is a Western terminology, its approximation in Chinese strategic thinking is the PLA's ASCEL. Nowadays China's A2AD capability is even more lethal than two decades ago, including cyberwarfare, anti-satellite weapons, carrier-killer missiles and the new-generation ships, planes, and drones. Cyber attacks are new dimension of the future warfare and the target is the information infrastructure of an adversary. Military and civilian cyberwarriors will be the tooth and the tail or frontline and support relationship in warfare. One of characteristics of the A2AD strategy is the wide proliferation of ballistic and cruise missile technologies and the convergence of Chinese military power around a missile-centric, rather than the conventional platform-centric, model of mass-firepower combat. The threats of the PLA's A2AD cover not only Taiwan, but also U.S. military bases on Japan. In addition to attacking land-based targets, the PLA has spared no efforts to develop ASBM capability to hit American aircraft carriers, though it is still a tough task.

The passage of PLAN surface warships and submarines through the Ryukyu island chain also demonstrates the penetration by China's A2AD threats of the first island chain. The year of 2008 is a watershed that the PLAN transformed from a brown-water navy to a blue-water navy. Prior to 2008, there were few PLAN activities in the Ryukyu region. Since 2008, particularly in 2013, the PLA claims it has become a regular training practice. The PLA's increasing Ryukyu activities have shown its strategic intents. Beijing has pledged to defend its land and territorial waters, including the disputed Diaoyu Islands. The maritime rights it desires to protect involve access to sea lanes, including the vital straits that transit the Ryukyus to the Pacific. Finally, should China resort to force to prevent Taiwanese independence, seizing the Diaoyu Islands or other islands in the Sakishima Chain would block possible U.S. or Japanese intervention.

Beijing has bolstered anti-access weaponry to pit against its superior rival. China's enhancing its A2AD capability has not only cast serious impacts and repercussions in the Asia-Pacific region, but also undermined goodwill and confidence-building in the cross-strait relations. One simple option for the U.S. and its partner nations to counter the PLA's increasing military strength is to boost their own version of A2AD capability. The items for Taiwan's anti access arsenal include supersonic ASMs, missile patrol boats, land-attack ballistic and cruise missiles, submarines, etc. Also, Taiwan should reshape its force structure with a strategy of asymmetry, combat credibility and resilience, as well as cooperating with the U.S. Asia pivot. Taiwan needs self-help and a robust defense and makes contributions to the peace and stability in this region.

China's Military Power Projection in 2020s: Intentions, Capabilities, and Constraints

Wei Hwa Chen
(Associate Professor of Graduate School of International Affairs, Ming Chuan University Taiwan)

Abstract

As China is involving more in regional disputes, so is stronger its military power buildup. The PLA stands nowadays on the threshold of major changes that may affect stability not only in the Asia-Pacific region but also serious implications to the interests of the U.S. The study discovers that there are several possible scenarios for military expansion in a future succession. Initially, such involvement would probably take the form of joint military operation by building aircraft carrier battle groups which have long been considered as major instrument for sea power acquisition. Next, building highly modernized sixth generation war planes, J-20 for instance, would be the next step for the involvement. Lastly, precise configuration of forces will be largely determined by the intensity of possible conflict with neighboring states. Hence the process of military modernization of PLA seems highly dynamic, and evaluating it will require constant acquire accurate data to be applied to the existing framework of understanding China's military muscle. Although China's defense budget has been growing at a steady pace over the past two decades, military experts have said the country's military spending is still far from the level it needs to be, as the country faces increasingly severe security challenges.

Ⅰ. Introduction: Significance of 2020s

On 2014 National People's Congress Annual Sessions, China announced that it would increase its military budget for 2014 to almost $148 billion, a 12.2 percent rise over last year—two decades double digits, except for 2010, in military expenditure. Fu Ying, spokeswoman for the second session of the National People's Congress 2014, however, argues that a country's military power should be viewed in terms of its policy trends, rather than merely military figures. She also contends that "if China becomes weak would regional peace could better be maintained?" Furthermore she insists that "peace can only be maintained by strength."[1]Xinhua News Agency also stresses that although China's defense budget has been growing at a steady pace over the past four years, Chinese military experts have said the country's military spending is still far from the level it needs to be, as the country faces increasingly severe security challenges.[2]

Despite China's justification its legitimacy over the military buildup by official rhetoric, skeptics of China's military spending claim that the rapid growth in defense budget is a clear sign of the country's goal of becoming a dominant military presence in the Pacific, with a navy and air force able to project power across the region. In fact, China has been perceived as a rising threat since the 1990s, and the threat theory has been created shortly after the start of the new millennium.

Indeed, the rapid pace and grand scale of China's rise have produced a heady mixture of wonder in the West. Many policymakers and military analysts have raised concerns over the potential for People's Republic of China (PRC) to mount a serious strategic challenge to both neighboring states and

---

[1] "No need for nerves over China's defense spending," *People's Daily*, March 5, 2014, <http://english.peopledaily.com.cn/90785/8556259.html>.

[2] "No need for nerves over China's defense spending," *Beijing Xinhua News Agency*, March 5, 2014, <http://news.xinhuanet.com/english/special/2014-03/05/c_133163698.html>.

the United States, especially in the East China Sea and South China Sea, sometime in the next decades. These concerns are based more on China's growing military expanding than on economic development.

Some even predict that these concerns may become reality by 2020s.[3] Hu Angong, a research fellow at Brooking Institute, argues that if current trends continue, China may become "A New Type of Superpower" in 2020s.[4] Others believe that China will be tempted to use its power projection capabilities to advance its distant interests in the current era, an ambition to restore its "traditional" glory as the leading power in Asia.[5]

Indeed, much of recent discussion about "China threat" for today and tomorrow focuses on the question of what intentions and strategies are behind the military restructure and modernization of the People's Liberation Army. As the use of force for the Chinese Communist Party (CCP) to maintain power has long been a crucial policy instrument since 1920s to the preservation of the regime today, [6] so it goes without saying that CCP may use power to extend China's past glory for the sake of its own survival.

So, is China an aggressive, expansionist power with enough resources to overwhelm regional states in order to create a new "Chinese Empire" in 2020s? Or is China still a vulnerable power in the face of numerous security challenges in spite of its military modernization? As a matter of fact, there are two contending views on China's rise, the realist one that perceives it as a threat to the existing order, and the liberal one that concentrates on the role

---

[3] Keith Crane et al., *Modernizing China's Military: Opportunities and Constraints* (Santa Monica, CA: RAND Corporation, 2005), p. xv.

[4] See Hu Angang, *China in 2020: A New Type of Superpower* (Washington DC: Brooking Institute Press, 2011).

[5] Richard D. Fisher Jr., *China's Military Modernization: Building for Regional and Global Reach* (Westport: Praeger International, 2008), pp. 169-170.

[6] Ka Po Ng, *Interpreting China's Military Power: Doctrine makes readiness* (London: Frank Cass, 2005), p. 8

of China integrating into the broader international community. This paper, judged from its title, will obviously focus more on realist perspective than on neoliberals that claim that conflicts could be mitigated through a closer economic cooperation.

Moreover, the approach of this paper is to analyze publications of mainland China's military as well as books and periodicals previously published Western analysis of China's military modernization, through which author intends to employ these findings in the event of PRC's armed forces possible future options. By doing so, this paper aims to make sense of PLA's power projection in 2020s by applying two major concepts to examine the purposes and capabilities of its military muscle; namely, military readiness and military strategy. Finally the author would conclude by offering some implications, based on predicting the PLA's capabilities in 2020s, to both regional states and the world. There are some limitations the author has to admit that any attempt to understand China's military strength and strategy faces the problem of the lack of transparency. Possible error and disinformation quite often than not occur. In this regard, documentary resources, books and essays related to the PLA development need to be scrutinized.

II . PRC's Intentions

In July 2012 China tried a new strategy by declaring that most of the 3.5 million square kilometers South China Sea had become Sansha, the latest Chinese city. The area China claims is as part of Sansha comprises over two million square kilometers of largely open ocean and a few hundred tiny islands and reefs, many of which are only above water during low tide, which is administered from one of the Paracel islands (Woody Island). The U.S. government and most regional states responded by asking China to obey international law in relation with territorial waters and the EEZ (Economic Exclusive Zone). China has not backed down apparently, but did not become aggressive again until November 23, 2013 when China claimed control over large areas of international air space via an expanded ADIZ (Air Defense Identification Zone). China wants all military and commercial aircrafts in these new ADIZs to ask permission from China before entering. Local states responded by sending in military aircraft without telling China, but warning their commercial aircraft operators to cooperate because it is considered impractical to provide military air cover for all the commercial traffic. China sees this as a victory, despite the obvious intention of other states to continue sending military aircraft through the ADIZ unannounced and despite whatever threats China makes. In response to that China has begun running combat air patrols through the ADIZ and apparently intends to try to intimidate some of the smaller countries defying the ADIZ.

For quite a long time, China has been demonized as a warmonger because international media and military analysts are fixated on portraying China as a threat to regional as well as global peace. In the face of external sceptic on China's role played in the international community, leaders of China in fact internally have confronted two contending views on its presence on the world stage. Some have supported the guidance from former paramount leader Deng Hsiao –Ping's "hide and bide" strategy in order to maximize its internal stability and external commercial interests so that a comprehensive national power could thus be accumulated. This is particularly major policy implementation before Hu-Jin-tao Administration.

For instance, Dai Bingguo, a former State Councilor of PRC and supporter of Deng Hsiao-ping, insisted that China should adhere to policy of "path of peaceful development and would not seek expansion or hegemony."[7] Such a strategy had also been implemented by former Chinese leader

---

[7] Dai Bingguo, "Adhere to the Path of Peaceful Development", *Ministry of Foreign Affairs, the People's Republic of China*, December 6, 2010, as cited in USC U.S.-China Institute Website, <http://china.usc.edu/ShowArticle.aspx?articleID=2325>.

Hu Jin-tao to rebut against the allegation of "China threat theory."[8]The view seeks to characterize China as a responsible world leader, emphasizes soft power, and vows that China is committed to its own internal issues and improving the welfare of its own people before interfering with world affairs. The term suggests that China has been seeking to avoid unnecessary international confrontation during the period of national power establishment.

Others, however, asserted that China would be best treated by a firm stance in the face of outside challenges, both from the United States and regional states. This could be easily identified ever since Xi Jing-pin took power.

As a newly appointed leader of China in 2013, Xi Jing-pin has frequently pointed out his "China Dream" and intended to apply the principles of "new type great power relations", *no conflict or confrontation, mutual respect, and win-win cooperation*, to its relationships with other major powers, the U.S. in particular, which implies a status commensurate with its growing economic and military power. Xi's three points indicate a strategy of *no zero-sum game, a pattern of opposing each other coming into conflict* and *beneficial way of peaceful co-existence.*[9] Apparently, Xi's words reflect China's expectation to be a traditional power at the world stage.

For years leaders of China have characterized the first two decades of the 21st century as a "strategic window of opportunities." and they routinely emphasized the goal of reaching critical economic and military benchmarks

---

[8] Sujian Guo, "Introduction: Challenges and Opportunities for China's 'Peaceful Rise'," Sujian Guo, ed., *China's Peaceful Rise in the 21st Century: Domestic and International Conditions* (Hampshire: Ashgate Publishing, 2006), pp. 2-3.

[9] Ren Xiao, "Modeling a 'New Type of Great Power Relations': A Chinese Viewpoint," *Open Forum*, October 4, 2013, <http://www.theasanforum.org>; See also in Rudy de Lon and Yang Jiemian, et al., "U.S.-China Relations: Toward a New Model of Major Power Relationship," *Center for American Progress*, February 2014, pp. 5-6, <http://www.americanprogress.org/wp-content/uploads/2014/02/ChinaReport-Full.pdf>.

by 2020.[10] In their view modernization of military strength is essential for China to achieve great power status. Officially, Chinese leaders claimed that increasing the scope of PRC's military capabilities is intended to build capacity for international peacekeeping, humanitarian assistance, disaster relief, and protect sea lanes.[11] Despite China's desire to better its image of developing peaceful development, its efforts to protect sovereignty and territorial integrity by strengthening military power in fact have frequently manifested in assertive rhetoric and behavior that incur regional concerns about its intentions and possible threat to the neighboring countries. In this respect, an assessment of threat perception has become an important analytical tool for knowing China's security concerns.

To be precise, an evaluation of PRC's threat perceptions usually starts by understanding the PLA's conception of China's national security and national interest. PRC leaders as well as PLA strategists consistently emphasize the need to maintain the existence of core national interests for China to survive and prosper. In order of importance, Beijing has officially defined its "core interests" as uncompromising three components: (1) "state sovereignty, national security, territorial integrity and national reunification"; (2) "China's political system established by the Constitution and overall social stability"; and (3) "the basic safeguards for ensuring sustainable economic and social development".[12]

---

[10] Quoted from Annual Report to the Congress, *Military and Security Developments of Involving the People's Republic of China 2013*, Office of the Secretary of Defense, p. 15, See also Dai Bingguo, "Stick to the Path, of Peaceful Development," *Beijing Review*, December 21, 2010.

[11] Information Office of the State Council, *China's Defense White Paper* 2013 (The Diversified Employment of China's Armed Forces, Chapter V: Safeguarding World Peace and Regional Stability), April 16, 2013.

[12] Information Office of the State Council, People's Republic of China, "China's Peaceful Development," September 2011, *white paper*, <http://english.gov.cn/official/2011-09/06/content_1941354.htm>.

These themes have been explicitly and openly addressed at many official occasions, including in most recent 2013 *National Defense White Paper*. These security perceptions suggest a range of some specific hauntings. First and foremost, historical memory motivates political intention and political assertion justifies military expansion. China, without a doubt, has long been haunted by 150 years of humiliation from foreign power's invasion which would enhance forcefully its aggressive behavior. Besides, as some "imagined territories", such as Diaoyu Island (Senkaku), Spratlys, South China Sea and Taiwan, are still taken by the foreign hands, which China sees as undisputedly an internal issue, the PLA's long term goals therefore would have to fulfill the mission of recovery those missing territories,[13] meaning that the issue is non-negotiable and is on a par with other sovereignty issues that could justify military intervention. Regardless of the difficulties that China may be confronted with when it comes developing a capability to resolve these territories with forces, the PLA will have to make it credible or will get the capabilities needed.

Another important factor for observation lies on defense budget of the PRC. Generally speaking, defense spending can also be viewed as a better indicator for explaining and predicting China's military ambition in 2020s. By using a linear approach toward understanding possible military expenditure of PLA in the future, researchers from RAND estimate that Chinese military spending is likely to rise to $185 billion by 2025.[14] Recently, China announced its military expenditure in 2014 is estimated $148 billion, going from $139.2 in 2013 and constituting the world's second-largest military

---

[13] Gary Li, "China's Military in 2020," Kerry Brown, ed., *China 2020: The Next Decade for the People's Republic of China* (Cambridge: Woodhead Publishing, 2011), pp. 104-107; see also in Andrew J. Nathan and Andrew Scobell, *China's Search for Security* (New York: Columbia University Press, 2012), Chp. 8.

[14] Crane, et al., *Modernizing China's Military: Opportunities and Constraints*, pp. xxiii-xxVI.

budge.[15] In reality, China's official annual defense budget now has increased for 22 consecutive years and more than doubled since 2006. Obviously, the figure RAND predicted on China's military budget in 2020s could possibly be underestimated if the trend continues, a double-digit increasing annually. However, there is also a problem for the analysis because different institutions such as CIA the World Factbook, US DoD's Defense Intelligence Agency (DIA), RAND Corporation, International Institute for Strategic Studies' (IISS) Military Balance, and Stockholm International Peace Research Institute's (SIPRI) Yearbook, may offer different figures for the assessment of China's defense budget.

A third factor for an evaluation of PRC's intention focuses on PLA's military policy and grand strategy. As noted earlier, armed forces of China are considered tools of CCP. Any military activity and strategy conducted can easily be attributed to the political purposes behind. However, there are some implications can be traced from Chinese military specialists. For instance, Liu Yazhou, a general and political commissar at the PLA National Defense University, warned "An army that fails to achieve victory is nothing." He pointed out that armed conflicts over China's maritime disputes would be a good chance to test the PLA's military prowess and "Those borders where our army has won victories are more peaceful and stable, but those where we were too timid have more disputes."[16] In other words, PLA military strategy still follows the instruction of political purposes. Hence, there is no denying the fact that without reliable military muscle as a backup political intentions are hard to be carried out.

---

[15] Shannon Tiezzi "China's Growing defense Budget: not as scary as you think, " *The Diplomat*, February 05, 2014; "World military spending falls, but China, Russia's spending rises," *Stockholm International Peace Research Institute*, April 15, 2013.

[16] Quoted from Minnie Chan" Fighting in East, South China Seas would test PLA prowess, general says," *South China Morning Post*, Feb 16, 2014, <http://www.scmp.com/news/china/article/1406458/fighting-east-south-china-seas-would-test-pla-prowess-general-says>.

III. Capabilities

As mentioned above, China's armed forces constitutes the world's second-largest military budge, yet is the largest military force in the world with a strength of approximately 2.3 million active troops, commanded by the Central Military Commission (CMC). The armed forces of PRC consist of a strategic force—the Second Artillery Corps, the Air Force (PLAAF), the Navy (PLAN),and Ground Force (PLAGF). In addition to the four main service branches, the PLA is supported by two paramilitary organizations: the People's Armed Police and the People's Liberation Army Milita.

China's military modernization started right after seeing highly advanced military technology of the U.S. conducting in the Gulf War in the 1990s. Noticing its backwardness in military strength and technology in comparison with major powers in the world, the PLA decided to adopt a leapfrog development of military modernization by accelerating military technological informationalisation while undergoing mechanization in order to narrow gap between its military forces and modernized forces of the world, particularly the Unite States.[17] Even though PLA analysts admit the technological weakness and learn from the war examples, they suggest a strategy of boosting the overall military capability.[18]

1. Strategic Forces (The Second Artillery Corp)

The Second Artillery Force of China is a strategic force under the direct command and control of the CMC, and the core force of China for strategic deterrence. It is mainly responsible for deterring potential enemy from using nuclear weapons against China, and for conducting nuclear counterattacks and precision strikes with conventional missiles. Tang Jiaxuan, a former

---

[17] Bi Wenbo, "Historical Consideration on Military Affairs," *China Military Science*, Vol. 2 (2002), pp. 75-82

[18] Fu Quanyou, "The World-wide Revolution in Military Affairs," *China Military Science*, Vol. 2 (2002), p. 9

State Councilor and Minister of Foreign Affairs, indicated that China recognizes its rising global position but wants to avoid major international attention or conflict through at least 2020. Thus, China's grand strategy through 2020 forms an important basis for more specific military objectives. As previously stated, Beijing is pursuing a comprehensive force modernization effort over the course of the first half of the 21st century.[19]

Though officially China appears to adhere to a doctrine of minimum deterrence, there is evidence to suggest that in recent decades China has moved or is moving to a limited deterrence nuclear doctrine. Some China analysts, Phillip C. Saunders & Jing-dong Yuan for instances, pointed out at "The Testimony before the U.S.-China Economic and Security Review Commission, Hearing on —Developments in China's Cyber and Nuclear Capabilities" that China's nuclear strategy has been changing as its strategic forces are modernized. China used to highlight "Minimum Deterrence", intended to enhance the survivability and effectiveness of its strategic nuclear force. Now, a doctrinal shift from minimum to "Limited Deterrence" could be triggered when China's new generation of JL-2, DF-5A, DF-31 and DF-41 ICBMs with multiple warheads are deployed in 2020s.[20] (SeeTable-1)

---

[19] Quoted from Office of the Secretary of Defense, *Military Power of the People's Republic of China 2009*, p. 3

[20] Phillip C. Saunders and Jing-dong Yuan, "China's Strategic Forces Modernization: Issues and Implications," *Testimony before the U.S.-China Economic and Security Review Commission, Hearing on —Developments in China's Cyber and Nuclear Capabilities*, March 26, 2012, pp. 3-5; See also in Connor Forman, "China: A Threat Assessment Through the Lens of Strategic Missiles," *Global Security Studies*, Vol. 1, Issue. 3 (Fall 2010), p. 5.

Table-1: China's Strategic Forces

Years / Missiles		2013	2020	2025	2030~
ICBMs	Type	DF-5A, DF-31, DF-41,	DF-5A, DF-31, DF-41	DF-5A, DF-31, DF-41
	Number	150+ (Road Mobile)	200-300 (Road Mobile)	300-500 (Road Mobile)
	Range (km)	13,000+	13,000+	13,000+
MRBMs	Type	DF-3/3A, DF-4, DF-21/DF-21A, JL-1, JL-2	DF-3/3A, DF-4, DF-21/DF-21A, JL-1, JL-2	DF-3/3A, DF-21/DF-21A, JL-1, JL-2
	Number	100	100+	100-300
	Range (km)	2,800+ 8,000+ (JL-2)	2,800+ 8,000+ (JL-2)	2,800+ 8,000+ (JL-2)
SRBMs	Type	DF-11, DF-12, DF-15	DF-11, DF-12, DF-15	DF-11, DF-12, DF-15
	Number	2,000+	2,000-3,000+	2,500-3,000
	Range	280+	280+	280+

Sources: Phillip C. Saunders & Jing-dong Yuan, "China's Strategic force Modern-ization and Implications for the United States," Testimony before the U.S.-China Economic and Security Review Commission, Hearing on —Developments in China's Cyber and Nuclear Capabilities, March 26, 2012, p. 7.

Mark Stoke's research indicates a possible development of China's missile capability, including 2015: 3000km range ASBM, 2020: 8000km range ICBM precision strike capability, and 2025: global precision strike capability.[21] Another China expert, Roger Cliff, in his 2010 testified in front of the U.S.-China Economic and Security Review Commission, also made some interesting predictions. He forecasted that by 2020 the weaponry of China's military forces will be roughly comparable to that of the U.S. military in 2000. Anthony H. Cordesman and Martin Kleiber shared the same view by predicting that China has its 3-step strategy for the development of credible forces, including "solid foundation" in 2010, "major progress" by 2020, and "capable of winning informationalized war" in the 21st century.[22]

Clearly, experts could interpret this many different ways. One certain way to look at is that even in 2020s China's military forces will still be 20 years behind the U.S. military. But the question, however, is to ask how much more advanced the U.S. military will be in 2020 as compared to 2000.

It is, of course, impossible to forecast precisely how China's strategic arsenal will develop in the coming decades, but continued development of SRBMs, ASBMs, and ASCMs are likely to be the focus of China's efforts moving forward due to their capacity to serve regional strategic objectives, including regional coercion in combination with anti-access to ensure access to resources and deter U.S. intervention in territorial and Taiwan disputes. Given China's nuclear strategy of limited deterrence, it shows a dizzying array of internal and external political environments that are likely to exert

---

[21] Mark Stokes, "China's Evolving Conventional Strategic Strike Capability," *Project 2049 Institute*, September 14, 2009, p. 2 ; Mark Schneider, "The Nuclear Doctrine and Forces of the People's Republic of China," *Comparative Strategy*, Vol. 28, No. 3 (July/August 2009), pp. 244-270; Office of the Secretary of Defense, *Military Power of the People's Republic of China 2009.*

[22] Anthony H. Cordesman and Martin Kleibert, *Chinese Military Modernization : Force Development and Strategic Capabilities* (Washington DC: Center for Strategic & International Studies, 2007), pp. 56-60.

force on China's strategic planners, pulling them in different directions and many believe that the potential for greater PLA influence over nuclear doctrine may move China in the direction of nuclear war-fighting strategies and a larger nuclear arsenal.

In sum, missile intimidation is a core Second Artillery mission, and works to restrain the enemy's strategic attempts or important risky military actions. Despite China's effort to modernize and continues to increase in size and sophistication of its strategic forces, it is highly unlikely to seek numerical parity with the United States and Russia, even if the U.S. and Russian arsenals fall to numbers well below current level. It is therefore fair to say that China's limited development of nuclear weapons "will not compete in quantity" with the nuclear superpowers. Instead, Beijing intends to maintain the "lowest level" of strategic forces sufficient to safeguard its national security in the coming decades.

2. Air Forces

Basically, there is no PLA "air force" to be discussed individually, even though a clear main task of PLA air force is to protect homeland security. In fact, there are multiple air forces spread across the PLA military, including air and naval aviation, ballistic missiles and surface-to-air missiles. In other words, the PLA air force, unlike other branches of military sects, cannot be separated from any branch of military when it comes to mission implementation.

Looking back in late 1990s, the PLA inventory of fighters was estimated 3,200, mostly "4th generation" Su-27 Flankers and 1950s era MiG 19s and 21s. Ten years later, the picture is very different. PLAAF has made significant progress in modernizing its air force since the early 2000s by purchasing Russian fighter jets such as the Su-30& Su-35, and also manufacturing its own modern fighters, most notably the J-10 and, J-15 and J-16. The most significant development for the PLAAF has been a transition from a large force of outdated fighters to a smaller, more capable force. Several fighters brought into service in the 2000s, some were purchased from Russia, while others were built under license by China's two major combat aircraft manufacturers, Shenyang Aircraft Corp. and Chengdu Aircraft Industry Group. Together they total nearly 400 aircraft whose aerodynamic characteristics and armament may be close to par with US fighters, excepting the F-22.

China now builds a single-engine fighter, the J-10, that RAND says is equivalent to the F-16, and a twin engine heavy fighter, the J-11B, considered better than the F-15. A test flight of J-20 in January 2011, the fifth generation of fighter prototype with stealth attributes, improved avionics and supercruise-capable engines, highlighted PLA's ambition to advance its air offensive defense and expected to be in service by 2017-2019.[23]

Officially PLAAF has not released its concrete principal mission, many would assume that its primary mission was to carry out air defense and to protect its homeland security. Then in 2008 China Defense White Paper first time highlighted PLAAF's mission as "strategic service," suggesting a broader mission than in the past and a greater emphasis on offense.[24] Many believe that PLAAF now is in a transition from a focus on territorial defense toward an air force that increasingly emphasizes offensive missions and trying to seize the initiative in its combat mission.[25]

---

[23] Quoted from *The Institute for national Security Studies (INSS)*:
http://www.defense-update.com/products/j/29122010_j-20.html

[24] Information Office of the State Council, *China's Defense White Paper*, 2008, p. 45.

[25] Murray Scot Tanner, "The Mission of People's Liberation Army Air Force," Richard P. Hallion, Roger Cliff, and Philip C. Saunders, eds., *The Chinese Air Force: Evolving Concepts, Roles, and Capabilities* (Washington DC.: INSS, 2012), p. 133.

Table-2 Assessment of PLAAF Combat Forces 2010-2015

Type	Name	Produced by	Number	Status
Fighter	J-7, J-8	PRC	860	retired
Fighter	J-10, J-10s, J-10B	PRC	350+	On duty
Fighter	J-11, J-11B, J-11Bs	PRC	280+	
Fighter (carrier-capable	J-15, J-15s (test fly)	PRC	12	
Fighter	J-20, J-31 (test fly)	PRC	unknown	
Fighter	SU-27SK, SU-27UBK	Russia	59+	retiring
Fighter-bomber	J-16	PRC	24+	
Fighter-bomber	J-7A, J-7B	PRC	300+	
Bomber	SU-30	Russia	73 24 (PLAN)	
AWACS	E-200	PRC	1	
AWACS	E-500	PRC	test fly	

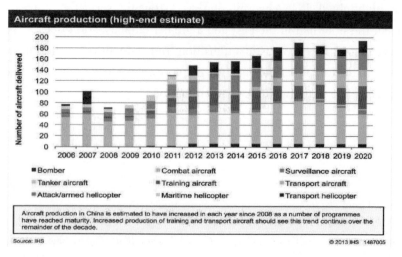

Source from:" Aiming high: China's air ambitions, "*IHS Jane's Defence Weekly*, 2013, <http://www.janes.com/magazines/ihs-janes-defence-weekly>.

Others believe that a major transformation of PLAAF missions in strategic thinking has been its growing faith in power projection capabilities that could be an important tool for CCP's political utility. In this respect, preventing Taiwan from declaring independence along with interest of safeguarding territorial integrity in the disputed area are presumably the driving forces.[26]

In spite of high achievement in improving its aircrafts technology and war-fighting functions, analysts tend to believe that China still depends very heavily on Russia and Ukraine in high performance turbofan engines right now, one that usually defines mission implementation. Besides, China has invested large amount of resources in the 99M project, which is still lagging in progress to the 117S project used on Su-35. If Russia is having trouble developing and mass producing a new generation of engine for aircraft, then one can imagine the stumbling blocks facing next generation Chinese turbofan projects. However, China is ahead of Russia in electronics and are ready to cost efficiently produce the components needed for radar system, whereas Russia is not. Up to now, all indications are that China's aircraft modernization development, in general, is going well. The first public flight in January 2011 of a stealthy new Chinese fighter, the J–20, came as a surprise to many observers who had agreed with then-Secretary of Defense Robert Gates that China would "have no fifth-generation aircraft by 2020" and only "a handful" by 2025.[27] Underestimation of China's air power would thus appear.

In sum, the challenge for PLAA in the next decade will not be developing new indigenous combat aircraft, but in closing the gaps that still exist. Nevertheless, given the political will and vast investments being made in the aerospace sector, it certainly appears to be a matter of when, rather than if,

---

[26] Murray Scot Tanner, "The Mission of People's Liberation Army Air Force," p. 138.

[27] Bill Sweetman, "Chinese J–20 Stealth Fighter In Taxi Tests," *Aviationweek.com*, January 1, 2011, <www.aviationweek.com/aw/generic/story.jsp?id=news/awst/2011/01/03/AW_01_03_2011_p18-279564.xml&channel=defense>.

this state of full maturity will be achieved. Whether or not the PLAAF can close the remaining gaps between its capabilities and those of the most advanced air forces remains to be seen. But one thing for sure is that the steps we have seen them taking are significant and deserve very close watching going forward.

3. Naval Forces

In the naval side, there are some developments of PLAN deserved special attentions here due largely to relevance to the PRC's power projection, particularly to both Taiwan and disputed territories mentioned before. According to U.S. China military analysts, Andrew Erickson and Gabe Collins, they predicted that if the trend proceeds PLAN will likely be the second large warships in number by year 2020, barring a U.S. naval renaissance and more importantly, China will possibly become the world's leading military shipbuilder in terms of numbers of submarines (See Table 3), surface combatants and other naval surface vessels produced per year with respect to overall shipbuilding. Moreover, by the assistance of Russian technical-proficiency levels PRC may even reach 2013 U.S. technical levels by 2030.[28]

Table-3 PLAN Submarine Order-of-Battle 2000-2020

Type \ Year	2000	2005	2010	2015	2020
Diesel	60	51	54	57-62	59-64
Nuclear	5	6	6	6-8	6-9
Ballistic	1	2	3	3-5	4-5
Total	66	59	63	66-75	69-78

Source: Craig Murray, Andrew Berglund, and Kimberly Hsu, "China's Naval Modernization and Implications for the United States," *U.S.-China Economic and Security Review Commission*, August 26, 2013, p.6.

---

[28] Craig Murray, Andrew Berglund, and Kimberly Hsu, "China's Naval Modernization and Implications for the United States," *U.S.-China Economic and Security Review Commission*, August 26, 2013, p.6.

As the world focused their eyes on the first aircraft carrier cruising in the South China Sea in recent days, the second carrier is being built at a shipyard in the port city of Dalian. According to the PLA official, construction is expected to be completed in six years, and China will eventually have at least four aircraft carriers. A report from the website of Hong Kong-based Ta Kung Pao newspaper, China will be building aircraft carriers with a lower tonnage which would only suffice for low-intensity battles. In fact, aircraft carrier development is core to the PLA navy, and could serve effectively as a deterrent to countries who provoke trouble at sea, against the backdrop of the U.S. pivot to Asia and growing territorial disputes in the South China Sea and the East China Sea.

As for naval power construction, since the 1990s, the PLAN has transformed from a large fleet of single mission platforms, coastal defense mainly, to a leaner force equipped with more modern, multi-mission platforms. In contrast to the fleet of just a decade ago, many PLAN combatants are equipped with advanced area air-defense systems, modern ASCMs (anti-ship cruise missile), and torpedoes. These capabilities not only increase the lethality of PLAN platforms, particularly in the area of anti-surface warfare in the region, but also enable them to operate beyond the range of land-based air cover.

Currently, the PLAN possesses some 75 plus principal surface combatants (destroyers and frigates), 50 submarines, 51 amphibious and medium landing ships, and 86 missile-equipped patrol craft (See Talbe-4). The PLA Navy has now completed construction of a major naval base at Yalong, on the southernmost tip of Hainan Island. The base is large enough to accommodate a mix of nuclear-powered attack and ballistic-missile submarines and advanced surface combatants, including aircraft carriers. Submarine tunnel facilities at the base could also enable deployments from this facility with reduced risk of detection.

Table-4 PLAN Naval Combat Forces

Type / Number	2006	2007	2008	2009	2010	2011	2012	2013
N-submarine	5	5	5	6	6	5	5	5
D-submarine	50	53	54	54	54	49	48	49
Destroyers	25	25	29	27	25	26	26	23
Frigates	45	47	45	48	49	53	53	52
Missile-armed Patrol ship	45	41	45	70	85	86	86	85
Amphibious LSTs & LPDs	25	25	26	27	27	27	28	29
Amphibious LSMs	25	25	28	28	28	28	23	36

Source: Table prepared by CRS based on data in 2000-2013 editions of annual
    DOD report to Congress on military and security developments involving
    China

As for China's aircraft carrier research and development program includes renovation of the KUZNETSOV-class aircraft carrier Hull 2 (formerly the Varyag), which began sea trials in 2011. It initially served as a training and evaluation platform. Once China deploys aircraft capable of operating from a carrier, it could offer a limited capability for carrier-based air operations. Now China has a land-based training program for carrier pilots and apparently it will still take several additional years for China to achieve a minimal level of combat capability for its aircraft carriers.

Besides, the PLAN is improving its long-range surveillance capability with sky-wave and surface wave over-the-horizon (OTH) radars. In combination with early-warning aircraft, unmanned aerial vehicles (UAVs), and other surveillance and reconnaissance equipment, the radars allow China to carry out surveillance and reconnaissance missions over the western Pacific. These systems can be used in conjunction with reconnaissance satellites to spot targets at great distances from Chinese territory, thereby supporting long-range precision strikes, including employment of ASBMs.

As for PLA's underwater fighting force, China is producing a new class of nuclear-powered ballistic missile submarine (SSBN). The JIN-class SSBN (Type-094) will eventually carry the JL-2 submarine-launched ballistic missile with an estimated range of some 8,000km. The JIN-class SSBN and the JL-2 will give the PLAN its first credible sea-based nuclear capability. Apart from that, China has expanded its force of nuclear-powered attack submarines (SSN). Two second generation SHANG-class (Type-093) SSNs are now in service and as many as five third generation SSNs will be included in the coming years. Other analysts believe the Type 093 SSN design will be succeeded by a newer SSN design called the Type 095 in 2015.[29]

When complete, the new class of SSNs will incorporate better quieting technology, improving China's capability to conduct a range of missions from surveillance to the interdiction of surface vessels with torpedoes and ASCMs. The current mainstay of modern diesel powered attack submarines (SS) in the PLAN submarine force are the 13 SONG-class (Type-039) units. Each can carry the YJ-82 ASCM. The follow-on to the SONG is the YU-AN-class (a Type-039 variant), as many as four of which are already in service. The YUAN-class probably includes an air-independent power system. The SONG, YUAN, SHANG and the still-to-be-deployed new SSN-class all will eventually be capable of launching a new long-range ASCM. China has deployed approximately 60 of its HOUBEI-class (Type-022) wave-piercing catamaran-hull guided missile patrol craft. Each boat can carry up to eight YJ-83 ASCMs. These boats have increased the PLA Navy's littoral warfare capabilities.[30]

Besides, PLAN has acquired modern, domestically-produced surface combatants. These include at least two LUYANG II-class (Type-052C)

---

[29] Ronald O'Rourke, *China Naval Modernization: Implications for U.S. Navy Capabilities—Background and Issues for Congress*, Congressional Research Service, February 28, pp. 7-14.

[30] Ibid.

guided missile destroyers (DDG) fitted with the indigenous HHQ-9 long-range SAM, with additional hulls under construction; two LU-ZHOU-class (Type-051C) DDGs equipped with the Russian SA-N-20 long-range SAM; and at least nine JIANGKAI II-class (Type-054A) guided-missile frigates, fitted with the medium-rangeHHQ-16 vertically launched SAM. These ships improve the PLA Navy's area air defense capability significantly, which will be critical as the PLA Navy expands its operations into areas beyond the range of shore-based air defense.

Table-5 PLAN Surface Orders-of-Battle 2000-2020

Type ⟍ Year	2000	2005	2010	2015	2020
Aircraft Carriers				1	1-4
Destroyers	21	25	25	28-32	30-34
Frigates	37	43	53	52-56	54-58
Corvettes				20-25	24-30
Amphibious Ships	60	43	55	53-55	50-55
Patrol (missile)	100	51	85	85	85
Total	218	162	218	219-234	224-247

Source: Craig Murray, Andrew Berglund, and Kimberly Hsu, "China's Naval Modernization and Implications for the United States," *U.S.-China Economic and Security Review Commission*, August 26, 2013, p. 7.

Although China's naval modernization effort has substantially improved it's naval capabilities in recent years, observers believe China's navy currently has limitations or weaknesses in several aspects in 2020s, including capabilities for sustained operations by larger formations in distant waters, joint operations with other branches of China's military, antisubmarine warfare (ASW), MCM, a dependence on foreign technology and suppliers for some ship components, and a lack of operational experience in combat situations.[31]

In sum, the sufficiency of PLA's naval capabilities is best assessed against that navy's intended missions. Although PLAN has its weakness and some limitations carrying out operations in-distant water, it may nevertheless still become sufficient to carry out wider array of tasks if Beijing has limited political intentions. In short, the PLAN is strengthening its ability to perform a range of regional missions. Over the next decade China may complete its transformation from a coastal navy to a navy capable of executing multiple missions in the region.

4. Ground Forces

The PLA has about 1.25 million ground force personnel, roughly 400,000 of whom are based in the three MRs (Military Region) opposite Taiwan. China continues to gradually modernize its large ground force. Much of the observed upgrade activity has occurred in units with the potential to be involved in a Taiwan contingency. Examples of ground unit modernization include the Type-99 third-generation main battle tank, a new-generation amphibious assault vehicle, and a series of multiple rocket launch systems. Approximately 40% of the ground force divisions and brigades are either armored or mechanized to deal with potential large scale conventional challenges, particularly border conflict.

In addition, the PLA ground force is composed of mobile operational units, border and coastal defense units, guard and garrison units, and is primarily responsible for military operations on land. The PLA mobile operational units include 18 combined corps, plus additional independent combined operational divisions (brigades), and have a total strength of 850,000. The combined corps, composed of divisions and brigades, are respectively

---

[31] Ronald O'Rourke, *China Naval Modernization: Implications for U.S. Navy Capabilities—Background and Issues for Congress*, p. 3.

under the seven military area commands (MACs): Shenyang (16th, 39th and 40th Combined Corps), Beijing (27th, 38th and 65th Combined Corps), Lanzhou (21st and 47th Combined Corps), Jinan (20th, 26th and 54th Combined Corps), Nanjing (1st, 12th and 31st Combined Corps), Guangzhou (41st and 42nd Combined Corps) and Chengdu (13th and 14th Combined Corps). In general, PLA ground forces are less likely involved in a territorial dispute with countries neighboring the South China Sea.

IV. Implications

1. Aim at Taiwan?

Military analysts tend to assume that PLA's short and mi-term goal for military modernization still focus on solving Taiwan issue. David Shear, for example, said in written answers to questions from the Senate Armed Services Committee, which is considering his nomination to be assistant secretary of defense for Asia and Pacific security affairs, adding that it's also necessary to understand what is shaping those investments. He pointed to work by the Defense Department's Minerva Initiative, designed to help assess future security challenges, which he said can help defense officials understand the social, cultural and historical factors driving China's strategic priorities. He stressed that China's increasing defense spending has been part of a long-term military modernization program lacking transparency but aimed at winning high-intensity, short-duration regional conflicts, primarily focused on Taiwan.[32]

---

[32]Nick Simeone, "DoD Asia Policy Nominee Encourage Close Watch on China," *U.S. Department of Defense*, Feb 25, 2014,
    <http://www.defense.gov/news/newsarticle.aspx?id=121721>; See also in Annual Report to Congress, *Military and Security Developments Involving the People's Republic of China 2011*, pp.55-61; David A. Shlapak, "The Red Rockets' Glare: Implications of Improvements in PRC Air and Missile Strike Capabilities," Roger Cliff, Phillip C. Saunders, and Scold Harrod, *New Opportunities and Challenges for Taiwan's Security* (Santa Monica, CA.: RAND & Institute for National Strategic Studies, 2011), pp. 73-78.

From a military standpoint, many argued that there is no significant strategic implication of China's Taiwan Strait scenario because without large scale production of amphibious vessels capable of crossing the channel poses no immediate threat to Taiwan's security.[33] In addition, Beijing would face great difficulty conducting such a campaign if leaders of PRC were willing to accept possible political, economic and military costs. Michael E. O'Hanlon and Richard C. Bush even predicted a possible defeat of China's military invasion. Nevertheless, PLA might succeed by barring third party intervention as its capabilities of strategic power projection proceed throughout the remainder of the decade. [34]

A different view shared by some, Zhang Baohui for instance, pointed out that the Taiwan Strait used to be a hotspot for military conflicts that could potentially drag U.S. into a major war with China. But now, the situation across the Taiwan Strait is totally different because of Taiwan's President Ma Ying-jeou's accommodation strategy toward Beijing, a new cross-strait relationship has emerged. Military tension and pernicious mutual mistrust have given way to institutionalized political dialogues, which would make a military conflict in the Taiwan Strait unlikely "even unthinkable."[35]

Indeed, the Taiwan Strait's security environment has been completely altered in the past several years. In his testimony before the House Armed Services Committee, Barry Watts, a senior research fellow of Center for Strategic and Budgetary Assessments, stated that since early 2000 the grow-

---

[33] Gabriel Collins and Lieutenant Commander Michael C. Grubb, "A Comprehensive Survey of China's Dynamic Shipbuilding Industry: Commercial Devlopment and Strategic Implication," *China Maritime Studies*, No. 1 (August 2008), pp. 41-42.

[34] Michael E. O'hanlon, "Why China Cannot Conquer Taiwan," *International Security*, Vol. 25, No. 2 (Fall 2000), pp. 51-86; Richard C. Bush and Michael E. O'hanlon, *A War Like No Other: The Truth About China's Challenge to America* (Hoboken NJ: Wiley and Sons, 2007), pp. 187-195.

[35] Zhang Baohui, "The Security Dilemma in the U.S.-China Military Space Relationship: The Prospects for Arms Control," *Asian Survey*, Vol. 51, No. 2 (April 2011), pp. 311-332.

ing migration of Taiwan's advanced technologies and businesses to mainland China, lured by such incentives as lower labor costs, have made direct military engagement least likely.[36]

The possible indications are that the gradual economic entanglement of Taiwan has continued, and that it is leading in the long run to Taiwan's eventual economic "capture" by the PRC. If this assessment is correct, then the chances of the PRC initiating a military takeover of Taiwan in 2020s appear to be quite low. Why use military force if economic entanglement leading to economic capture is succeeding? Note, too, that this approach embodies Sun Tzu's dictum that the acme of strategy is to subdue the enemy without fighting, Watts argues.[37] In this regard, keeping Taiwan in a "cage" would definitely serve more interest to China than a risky military attempt.

2. Regional Dominance?

At the ASEAN Defense Ministers Meeting Plus (ADMM+) in Hanoi in October, Chinese Defense Minister Liang Guang-lie responded calmly to former US Defense Secretary Robert Gate's reiteration of Hilary Clinton's ARF (ASEAN Regional Forum) comments on the South China Sea, opting to use the opportunity to reassure the region that China's military is not aimed at challenging or threatening anyone, and is defensive in nature.[38] Then why regional states still perceive increasing military threats from China?

---

[36] Barry Watts, Hearing Transcript of Congressional Testimony, "The Implications of China's Military and Civil Space Programs," Committee: Senate U.S.-China Economic and Security Review Commission, May 1, 2011, <http//www.uscc.gov/hearings/2011 hearings/written-testimonies//11-05-11-wrt/11>.

[37] Ibid.

[38]Thom Shanker, "US and China Soften Tone Over Disputed Seas," *New York Times*, October 12, 2010, <http://www.nytimes.com/2010/10/13/world/asia/13gates.html>.

For years, China's military sounded wary of its neighbors, naming Japan as a "trouble maker" while accusing the US of making situation "tenser" by forging alliances in Asia. Some analysts claim it has long been in Tokyo's interests to play up the "China Threat". But objectively speaking, the threat is real, and it becomes tangibly more worrying by the day," as a columnist contended.[39]

One different view argues that "China's military modernization makes perfect sense to me as a natural evolution commensurate with China's rise as a great power," says James Mulvenon, director of the Washington-based Center for Intelligence Research and Analysis. He believes that China will create an advanced and capable military to conduct and sustain at a greater distance from its border by 2020s.[40]

For the past two decades, China's military budget as percent of GDP has hovered at or close to the 2 percent line. This is not to say, however, that China's military build-up has posed major threat to destabilize the East and South-east Asian region. The growth in military expenses (and thus military capabilities) may be an entirely natural and expected result of China's rapid economic development, but it is still causing changes in the regional security environment. In fact, it will take some time for the Asia-Pacific to reach a new equilibrium that takes China's military into account. Worries about China's military expansion shouldn't be blown out of proportion by ignoring the surrounding context. Without a doubt, China's defense spending increase by billions each year, and that change in the balance of power can be dangerous. But the military build-up is also simply a natural part of China's economic growth. Therefore, the question remains if China would willingly decrease its military buildup for the sake of appeasing neighboring states.

---

[39]Simon Tisdall, "China's military presence is growing: Does a superpower collision loom?" *The Guardian*, January 1, 2014.

[40] Quoted from Jayshree Bajoria, "China's Military Power: Scope of the Threat," *Council on Foreign Relations*, February 4, 2009.

More importantly, will the Western Pacific region remain calm and stable if China decelerates its military budget? The answer seems uncertain, or even negative.

3. A Cold War Model ?

Hillary Clinton, the former US Secretary of State, highlighted the US policy in the west Pacific in her article "America's Pacific Century" in the **Foreign Policy** November 2011 issue, in which she frequently used the word "pivot" in outlining the evolving U.S. strategy toward the Asia-Pacific region. However the word was so enticing that the term "pivot to Asia" became iconic when discussing U.S. foreign policy toward the region. Many questioned the practicality and relevance of the word "pivot," rather than the actual implications of the policy under review. But the question is as China now rises and recasts the military and political dynamic in the region to reflect its own traditional centrality, will the US rebalance strategy become the defining geopolitical contest through a new era of military competition in the Pacific?

For some pessimistic IR scholar, John Mearsheimer for example, they argue that the US and China are doomed to repeat the intense security competition of the Cold War. He assumes that the rivalry could even more volatile than with the former Soviet Union because there are more potential disputes and he predicts China and Japan may "start shooting each other" at some stage over the next five years.[41] If, as Mearsheimer says, that is the case, the Western Pacific would turn into one of most dangerous flashpoints and the US would be forced to involve in the dispute. Then there will be a "Hot War" instead of "Cold War" scenario.

---

[41] John Measheimer, "Say Goodbye to Taiwan," *The National Interest*, March-April, 2014; See also in Geoff Dyer, "US vs China: is this the new cold war?" *FT Magazine*, February 20, 2014,
<http://www.ft.com/cms/s/2/78920b2e-99ba-11e3-91cd-00144feab7de.html#axzz2vdN2m0ao>.

The truth, however, is that since America's high-profile "pivot to Asia" strategy was announced, the Asian states worried about Obama Administration dramatic defense budget cut might thwart American strategy abroad, especially military presence in the Western Pacific region. Apparently the pivot aims to deter Chinese aggression and reassure U.S. friends by shifting 60% of U.S. naval forces to the Pacific by 2020, from recent 50%, and increasing cooperation among U.S. allies. Up to 2,500 U.S. Marines will now regularly rotate through northern Australia, and the U.S. Navy has begun basing littoral combat ships in Singapore. U.S. forces will likely gain greater bases access in the Philippines, possibly from Vietnam, and Washington is shoring up Japanese defenses with new radar and drones. But the problem, according to the ***Wall Street Journal*** report recently, is that since 2009 the Obama Administration has cut half a trillion dollars from defense. Now it wants to cut the Army to pre-World War II levels. It says it doesn't want to cut the Navy to World War I levels, but 30-year shipbuilding plans are expected to produce fewer than the minimum 306 ships the Navy says it needs to accomplish its missions. That will be a particular handicap in the maritime Asia-Pacific. Assuming the pivot proceeds, Asia in 2020s can expect to see 60% of a smaller U.S. Navy.[42] But, for China, that which most needs military transparency is the US. The country's return to Asia first of all means the return of its military presence. About 60 percent of its nuclear submarines, six of its 11 nuclear-powered aircraft carriers and troops withdrawn from Afghanistan will be deployed to the Asia-Pacific. The US has been claiming it is not targeting China. But is its real intention transparent?

Evidently, it is estimated, according to the DoD's QDR, that the U.S. military muscle will be too weak to deter potential adversaries if the US

---

[42]"America's Non-Pivot to Asia: US friends and foes know that entitlement spending inevitably draws from defense entitlement," *The Wall Street Journal*, March 9, 2014, <http://online.wsj.com/news/articles/SB10001424052702303369904579423530847513384>.

government keeps defense budget cut as planned, then it would be difficult for DoD to carry out its defense strategy by 2020s.[43]

Table 6 2001-2014 U.S.-China Defense Growth Rate (US billion）

Year	Budget		Growth rate (%)		Fiscal year (%)		GDP (%)	
	U.S.	China	U.S.	China	U.S.	China	U.S.	China
2001	366.7	21.10	1.73%	19.42%	3.61%	7.63%	2.9%	2.1%
2002	422.2	25.00	1.51%	18.43%	3.55%	7.74%	3.2%	2.2%
2003	484.2	27.93	1.47%	11.72%	3.58%	7.74%	3.6%	2.1%
2004	544.1	32.21	1.24%	15.31%	3.21%	7.72%	3.8%	2.1%
2005	601.2	36.23	1.15%	12.50%	3.11%	7.29%	3.8%	2.0%
2006	622.2	43.62	0.45%	20.40%	2.89%	7.37%	3.8%	2.1%
2007	653.9	52.04	0.51%	19.30%	2.95%	7.14%	3.8%	2.1%
2008	630.7	61,15	1.17%	17.52%	3.72%	7.70%	4.2%	2.0%
2009	675.2	70.68	0.88%	14.90%	3.55%	6.30%	4.6%	2.2%
2010	689.7	77.95	1.27%	7.50%	3.86%	6.30%	4.7%	2.1%
2011	708.8	91.52	1.07%	12.7%	4.01%	5.81%	4.6%	2.0 %
2012	671..0	106.4	−3.81%	11.2%	3.76%	5.20%	4.2%	2.0%
2013	613.9	139.2	−8.64%	10.7%	3.58%	5.15%	4.1%	2.0%
2014	574.9	148.8	−6.36%	12.2%	-------	------	------	------

Sources: US. Department of Defense Fiscal Year Supplemental Request, May
　　　2009, p. 3; See SIPRI, Yearbook, 2013; The World Bank,
　　　http://data.worldbank.org/indicator/MS.MIL.XPND.GD.ZS;
　　　<https://www.cia.gov/library/publications/the-world-factbook/rankor
　　　der/2034rank.html>;StateDepartment of China 2014,
　　　<https://www.cia.gov/library/publications/the-world-factbook/>.

From both political and economic perspectives, China's rise is inevitable and may contribute greatly to the peace, stability, and prosperity in the region. Yet, when focusing their eyes on China's military muscle, some

---

[43]"America's Non-Pivot to Asia: US friends and foes know that entitlement spending inevitably draws from defense entitlement," p. 56.

countries are still unable to balance themselves to this new situation. The U.S. is no exception. However, Bonnie S. Glaser, China specialist and a strategic analyst at CSIS, has noticed the different characters of US-Soviet and US-China's competition in nature. She concludes her findings at her recent research by suggesting that "Efforts will need to be made to ease Chinese suspicions about the US pivot to Asia without undermining deterrence. If Beijing were to conclude that the US was implementing a policy of strategic encirclement and containing China's rise, that could result in the adoption of a range of counter-containment policies that would be destabilizing to the region. On the other hand, a policy weighted in favor of providing strategic reassurance might be interpreted by Beijing as US weakness, and encourage China to be more assertive in the South China Se*a*."[44] In other words there is no simple formula for US-China relations. Actually "a cold war model" could do more harm than good, not only to the U.S. but also to the regional stability.

Undoubtedly, more assertive PLA by 2020s will signify a more confidence over disputed territories because of rapid increasing military capability and expansion along with high economic growth. In general, the PLANAF and PLAAF arms will be able to provide extensive air cover for both the naval and ground forces in 2020s in their carrying out abroad mission. However, any military expansion, according to Clausewitz, would not go without political constraints and its own limitations.

---

[44] Bonnie S. Glaser, "Understanding Recent Developments in US-China-ASEAN Relations: A US Perspective," *East Sea (South China Sea) Studies*, January 21, 2013, <http://nghiencuubiendong.vn/en/conferences-and-seminars-/hoi-thao-quoc-te-4/778-understand-ing-recent-developments-in-us-china-asean-relations-a-us-perspective-by-bonnie-s-glaser>.

V. Constraints

Whether the ongoing modernization of the PLA poses more of a threat to Taiwan or to the regional states, China's military strength improvements are still antiquated overall in comparison with the American forces. Apart from that, China's military modernization may face a lot of challenges now and in the following decades.

Firstly, it deserves to note that PRC's military, although getting stronger in its power projection to neighboring waters, is still decades behind U.S. military power in Asia. If the trend is not to develop into a war or an arms race in the near future, which the PRC has been increasing awareness of its role played in the region, Beijing needs to actively engage in more multilateral military confidence-building measures. Unfortunately, the growth of the region's military budgets reflects similar security concerns over China's intention.

Secondly, China's assertiveness abroad also belies growing insecurity at home. Richard D. Fish pointed out that CCP regime relies on the loyal support of the 2.25 million PLA members who are important tool of the communist party. In order to execute political control CCP need to pay increasing heed to strengthening military policies and modernizing armed forces so as to obtain greater global military influence. In other words, more military spending means more resources needed.[45] However, the obsession with avoiding the Soviet Union's fate could well prevent the CCP from carrying out reforms, both socially and economically. This "gun or butter" argument could thus been viewed as a touchstone for CCP, or China, to maintain in power. The real test of China willingness to keep military spending constant will come when China's headlong economic growth starts to slow further. Apart from that, one thing deserves special mention, according to BBC re-

---

[45]Richard D. Fisher Jr., *China's Military Modernization : Building for Regional and Global Reach* (Westport: Praeger International, 2008), p. 5

port, is that PRC's 2013 appropriation for maintaining public security, internal stability to be more specific, was up 10.8% (RMB 769,1 billion) from 2012 (RMB 701,8 billion), higher than 2013 defense budget (RMB 740,6 billion). Accordingly, the figure may be an indication of a getting more serious security concern to China's internal challenges than external threats.[46]r.

Thirdly, following the logic from previous point, the PLA has little time struggling for regional or global domination. Actually, common threats frequently highlighted by the Chinese government are terrorism, separatism and extremism. Uygur separatists in Xinjiang Province, for example, are constantly fomenting troubles. Besides, the country has 22,000km of land borders with 14 neighbor states, plus an 18,000km coastline. Safeguarding these borders is difficult and challenges, and it definitely occupies a large proportion of the military serviceman and resources.

Fourthly, China's military power projection in the region, particularly in the West Pacific, could confront the latest refocus of the U.S. so-called "Pivot to Asia", which will see 60% of American naval vessels deployed in the region by 2020, up from 50%, particularly the enhancement of critical naval presence in Japan and possibly in Southeast Asian states.[47] In this respect, the U.S. could effectively wage an allied strategy of rebalance in the region, even the U.S. government fiscal environment continues to deteriorate.

Fifthly, China is seeking to attain military strength commensurate with its growing regional and global influence and to close its technological gap with world major powers. To achieve this goal, China needs to import critical weapons and advanced technology it requires. By far, Russia remains

---

[46] Chen Chi-fen, "An Observation of China's Military Spending and Spending on Maintaining Stability," *BBC Chinese Website*, March 5, 2014,
<http://www.bbc.co.uk/zhongwen/trad/china/2014/03/140305_ana_china_npc_army.shtml>.

[47] U.S. Department of Defense, *Quadrennial Defense Review*, 2014, p. 34, 54; see also in Hannah Beech, "How China Sees the World," *Time*, June 17, 2013, pp. 26-27.

China's most important outside source of arms and technical assistance. The PLAN's best-known vessel - its sole aircraft carrier, the Liaoning - was purchased from Ukraine. From China's perspective, Europe interprets the arms embargo most generously, mostly blocking only lethal items or complete weapons systems. But for Europe, China is more an opportunity than a threat. Under Beijing's long-term policies to promote innovation, domestic arms makers are encouraged to import the foreign technology that China lacks. The challenge is to adapt this range of components and know-how into locally built weapon systems.

Finally, if China aims at increasing its influence in the region so as to restore its traditional centrality, it obviously requirs an alliance to balance possible challenges from U.S. and its existing allies. Alliance formation, according to Stephen M. Walt, is determined by the threat they perceive from other states and even a weak state is more likely to bandwagon with the rising threat in order to protect its security.[48] Under such circumstances, China would not be easy to get allies when it is considered more as a threat than a benign alliance.

If the United States is considered by PLA as main archrival in its regional expansion, then PLA would be to focus on developing anti-access (A2) and area-denial (AD) capabilities in order to prevent possible wartime intervention. It would likely require more aggressive patrols of its coast and airspace of its Exclusive Economic Zone (EEZ), both in East China Sea and South China Sea. Apparently, some of these activities are already occurring. However, such a more aggressive posture needs greater investment. To be prepared for carrying out A2/AD strategies, PLA needs to take some limited offensive power projections, including a combination of surface, ships, air-

---

[48] Stephen M. Walt, "Alliance Formation and Balance of World Power," *International Security*, Vol. 9, No. 4 (Spring 1985), pp. 3-43.

planes, and submarines operating far from their ports and bases, that would require a significant investment,[49] seemingly unlikely in the 2020s.

Currently, or in a near short-term, China does not possess significant military power strong enough to challenge the U.S. both in regional as well as in global arenas. Its military forces are inferior, and not by a small margin, to those of the United States. But power itself is rarely remaining static. The real question that is often overlooked is what happens in a future world in which the balance of power could be shifted sharply against Taiwan and the United States, in which China shall control much more relative power than it does today, and in which China is in roughly the same economic and military league as the United States in the next decade. In essence, one thing for sure is that a world in which China is much less constrained in 2020s than it is today.

VI. Conclusion: Future options

For China, power projection is political influence exerted at a distance through the use or threat of military force. It is useful to think of power projection operations according to their underlying political purposes. A sea change driven by political purpose apparently reflects on PLA's military modernization. Any change could take times but it is easy to forget that China's rise was what the West wanted. In the late 1960s, Richard Nixon, the former president of the United States, once stated before the public "We simply cannot afford to leave China forever outside the family of nations, there to nurture its fantasies, cherish its hates and threaten to its neighbors. There is no place in this small planet for a billion of its potentially most able people to live in angry isolation."[50] Since then most American presidents repeatedly said that the prosperity and stability of China are in the interest of

---

[49] Roger Cliff, Philip C. Saunders, and Scold Harold, *New Opportunities and Challenges for Taiwan's Security* (Santa Monica, CA: RAND, 2011), p. 121-122.

[50] Quoted from Nathan and Scobell, *China's Search for Security*, p. 1.

the United States and international community. Besides, many China watchers have predicted the collapse of the PRC for half a century only to be proved incorrect.

A pessimistic scholar of IR, John Measheimer for example, predicts that a powerful China isn't just a problem for Taiwan. It is a nightmare not only to regional states but also to the United States because an earlier hope that China would become a responsible stakeholder in the international community when it was rich turned out to be wrong.[51] Apparently, every power in order to guarantee its security and existence aims to maximize its power capabilities, not only in military terms but also in the economic and societal sphere. Realists discuss the issue of so-called latent power every state possesses and that represents a basis of its power status. Offensive realists, such as Measheimer, claim that it is necessary to achieve absolute power leading to hegemony in order to guarantee the state's position and security. Thus how the world sees China may matter just as much as China views the world.

If this is true to any power's rise as China, then why emergence of China as a power has always stirred an imagination of the Western World? As China's renaissance is by now a familiar narrative, so the story of its astonishing threat bears repeating. Clearly, China understands well that it is preparing to fight not one but different adversaries, as seen in the target-specific training. However, military muscle has to be interpreted. It does little good only to understand how big or how advanced an army is unless one knows what it is for, how it is to be used, and more importantly, how ready of its power is to implement what is intended. Obviously, the current PLA military force that is designed to make reflects a penchant for the "local war".

As noted earlier, forecasts are always difficult because of too many unknown factors interrupted. Both linear and non-linear approaches for future

---

[51] Measheimer, "Say Goodbye to Taiwan," *The National Interest*, March-April, 2014.

prediction have their own advantages and limitations. In spite of all these restraints some scenarios are likely to predict through files and historical experiences. Others might also become plausible largely due to a rapidly changing strategic environment.

Will China maintain its regional power status with global significance?[52]Or China sees itself as a regional power with an "Offensive-defense" (active defense) strategy?[53] These are questions with different answers. However, it is estimated from a linear approach that if the current trend of the PLA's modernization persists on another two decades, according to Lt General JS Bajwa's survey, PLA will be in a position to control the battle space up to 250 nautical miles off the eastern coast through denial strategy and joint operation at moderate scale.[54] If this is the case in the next decade, together with the factors mentioned previously, China's rise to great power status will certainly motivate the PLA flexing its muscles in 2020s as follows.

---

[52] Richard D. Fisher argued that China's military programs represent an aspiration to global, not just regional military power in 2020 to 2030, see Richard D. Fisher Jr., *China's Military Modernization: Building for Regional and Global Reach* (Westport: Praeger International, 2008), p. 4.

[53] KarlisNeretnieks, a senior research fellow at Institute for Security & Development Policy of Swedish National Defense College, characterizes China's defense policy as an "Offensive offense" strategy, meaning that China may conduct an advanced capability to launch limited, joint, mainly air and ground campaigners in areas bordering mainland China, see KarlisNeretnieks, " A Chinese Strategy for the 21st Century: An European View on China future Strategic goals and military development," paper presented at the" The Proceedings of Annual International Academic Lecture Series 2008" (Taiwan: ROC National Defense University, December 31, 2008), p. 10.

[54] JS Bajwa, *Modernization of the Chinese PLA: From Massed Militia to Force Projection* (Atlanta, GA: Lancer Publishers, 2013), Chp. 9.

1. Deploying more deterrent forces

A. Showing its military muscle so as to both intimidate potential rival and attract possible allies on a regional as well as global scale.

B. Relying more on nuclear-powered submarines so that a deterrent strategy could be implemented.

2. Unfurling its wings

A. Accelerating its transition from territorial air defense to both offensive and defensive operations.

B. Building robust capabilities in support of power projection, air lift, and close support to ground forces and naval forces in disputed territories.

3. Stretching the naval boundaries

A. Building a navy that could deny other navy access to and pose a certain threat to navies that operate in the Pacific and Indian Ocean.

B. Escorting its economic interest and protecting its energy resources acquirement.

C. Carrying out more long range military exercises in the disputed territories, East Asia Sea and South China Sea for instances.

D. Integrating a few aircraft carrier battle groups to enhance China's great power image as well as protecting its vital interest in disputed waters.

PLA Cyber warfare &Taiwan's information security

Ying Yu Lin ,林穎佑

(Assistant Professor,St.John's University)

Abstract

In recent years, information security is the hot topics. China almost certainly would mount a cyber-attack on U.S ,ROC, JP and other countries. As we know, C4ISR is the key point in modern warfare, internet is the corn about information wars .Although modernization of the PLA is a rising star , a serious threat to other countries, still can't confront US military force in front of them. So, Asymmetric warfare is the main thinking in PLA, either Anti – aircraft or "acupuncture warfare" (點穴戰). As IT advances, PLA combine the cyber-attack and Net Center warfare (NCW), cyber warfare is not only software but also hardware .PLA Cyber warfare (網電一體戰)is not just in cyber space, it's a part of information warfare.

Taiwan is the major target of PLA cyber army, APT attack (Advanced Persistent Threat, APT) , Botnet(僵屍網路), watering hole（水坑式攻擊）Social Engineering ,are total different form old attack style. For example, DDOS((distributed denial of service 分佈式阻斷服務攻擊) attack focus on hacker skill, but, now, strategy is the key point about APT attack. Security is a process not a product. Especially, the critical infrastructure protection is the keystone about homeland security. Although a lot of critical infrastructure protections (CIP 關鍵基礎設施) are nongovernment still uses the information system to control their operation, if hackers/cyber army invades the operation system, then CI would be out of control and will bring the panic of national secretly. In addition to the geostrategic is the traditional model to define PLA, island chain strategy also follows it. However, cyberspace is in the different track of thinking about war and beyond geostrategic. It's a new field about military warfare.

This study attempts to analyze the PLA cyber warfare, including their operation model and defense strategy. At first, define the cyber warfare in PLA and compare with cyber warfare and information warfare. Besides, attempts to discuss the content about cyber warfare and the common mode about cyber-attack from PLA, including the side of software and hardware. Finally, try to point out the way of defense cyber-attack and improvement information secretly in Taiwan. The information teleology is not the major issue in this study, Rootkits & system code won't discuss in this paper. This study will focus on the effect of attack model about PLA cyber army and point out the way to improve information security.

Key word

PLA, Information security, APT, Cyber Army

It was the best of times, it was the worst of times, it was the age of wisdom, it was the age of foolishness, it was the epoch of belief, it was the epoch of incredulity, it was the season of Light, it was the season of Darkness, it was the spring of hope, it was the winter of despair, we had everything before us, we had nothing before us, we were all going direct to Heaven, we were all going direct the other way- in short, the period was so far like the present period, that some of its noisiest authorities insisted on its being received, for good or for evil, in the superlative degree of comparison only.[1]

In the information world warfare consists of only cyber-attacks but no tank or missiles. Campaigns happens every day silently without informed by any government or media. In recent years, the media and government propaganda has exposed the general public to new cyber-warfare and information security-related issues. When America formally established the United States Cyber Command (U.S.CYBER.COM) in 2010, the cyber war officially went public. In 2009 U.S. government has published "National Cyberspace Policy Review", and this issue was raised to the "International Strategy for Cyberspace" by the White House in May of 2011. [2]

However, the army began in early stage in use of internet network technology to the war in the late 20th century which was considered to assist war efforts along with the arrival of the information technology era of combat In Gulf War, it's one ounce of silicon was more important than a ton of uranium. Military innovation would focus on how to efficiently use high-tech

---

[1] Charles Dickens, A Tale of Two Cities (London: Chapman & Hall, 1859), Chp. 1, Literature.org,

<http://www.literature.org/authors/dickens-charles/two-cities/book-01/chapter-01.html>.

[2] White House, "International strategy for cyberspace," May 2011,

<http://www.whitehouse.gov/sites/default/files/rss_viewer/international_strategy_for_cyberspace.pdf>.

methods to enhance combat effectiveness.[3] But in recent years certain concepts of terms and analysis have become unclear, which leads to misleading terms and phases in the cyber-war.

Ⅰ. Definition of Cyber-warfare and related glossary

1. Information Warfare and Network-Centric Warfare (NCW)

The three concept terms of "information warfare", "network-centric warfare", and cyber- warfare are the most commonly misused. Each of these terms has the own special meaning and different definitions; the imprecise wording of these terms with excessive use that - highlights a lack of professional understanding. Information warfare is a type of interference with enemy information, which are information processes, information systems, networks and security of information from the computer of our side, information processes, information systems, and computer networks to ensure our side's information has a competitive advantage. Therefore, information warfare is how to use information technology to assist warfare, and can be regarded as its own category.[4] Whether in decision-making or in directing combat, advances in technology have brought about revolutionary development in warfare. Information warfare should be referred as a concept of Instruments war, rather than the operations of themselves. So the key point about Information warfare is to forces on how to use the information power to win the war. It's an "approach" about war, not a type about war.

Information warfare is emphasizing on the integrated description of the military (application), and information technology (interface), but not the actual armed operations are insisted. In the PLA, the "information warfare" is a way to get the advantage of high-tech auxiliary warfare, including the use of computer technology to aid all tactics. Creating a network in each

---

[3] 艾文・托佛勒著，傅凌譯，《新戰爭論》（台北：時報文化出版社，1994 年），頁 90。

[4] 陳文政、蘇紫雲，《不完美戰場》（台北：時英出版社，2001 年 5 月），頁 19。

combat of units (tanks, ships, bases) were integrated to help each other. Unleashing their works to against enemy is a major key point in information warfare.

Network-centric warfare, (PLA is also known as 網路中心戰), which is the use of the information advantage so that each armed force can achieve information sharing (just in time) for joint operations and maximum fire in order to defeat the enemy in the shortest possible time.[5] The key being whether joint operations can achieve is integration information. This is especially true in the case of a limited budget, as network-centric warfare is the process of finding a way to effectively combine military units, C4ISR （Command、Control、Communication、computer、Intelligence、Surveillance、Reconnaissance）, and fire.[6]

2. Definition and Characteristics of Cyber warfare

In contrast cyber warfare is relatively easier to define. Cyber warfare refers to the use of the internet as a medium of attack, by using the information warfare concept for both a special form of offensive and defensive in cyber space. The characteristicof cyberspace is independent of time, geography, the impacts of weather, and it blurs traditional boundaries. It's totally defend other warfare. The opportunity of being used is increasing due to the lower cost threat range of cyber warfare. IP tracking of a specific user might be done by using of IT, however, proving the consistency of identification for the user is still technically difficult issue at present, especially in the registration of user's internet address and the real identity. Thus, as the internet develops, countries, companies, and individuals are all linked together through the internet. Since internet users can also hide their true identities, no one knows who the user is, and the difficulty is increasing because of

---

[5] Department of Defense, The Implementation of Network- Centric Warfare, 2004, p. I.

[6] 許秀影、劉豐豪、張瑞勇，《前瞻國軍對次世代網路之應用》（台北：國防大學管理學院，2010 年），頁 238-243。

maintaining information security and tracking.[7] What defines intrusion into a system by an individual hacker as criminal behavior, or a military operation into hostile enemy government organizations? Even if the source of the attack is determined, how can we clearly define whether it should be dealt with by a local security police system level or at the international level if it relates to homeland security? This leverages the difficulty to take any retaliation and counter-action steps.

Besides, electric data is deferent from the other goods, people often don't know that their computers were under attack. As we know, you will connect to the credit card company if your wallet was lost, and will contact with the system company once lose the cell-phone. However, you seldom figure out what you've lost on your desk. Generally speaking, it's just like to copy films from one side to another (hacker's) computer. Although the department of information security maybe can monitor it, one security bulletin can still destroy all system in the cyberspace. During the Cold War, technology limited information could be stolen in cameras or hidden in film by spies which took their risks of failure. But in the cyber space, anyone who can sit on a chair in the different faraway places and move through the internet network. They can easily download the data with they find. When the victim realizes that the data has been stolen, it tends to be a large amount of sensitive information. The major feature of cyber warfare is that it couldn't be able to detect problems, and another feature of cyber warfare is that victims do not know that their computer has been compromised. Access to electronic information is not the same as physical objects; while people lose wallets, watches, and passports those physical important things, it will be immediately noticed, but in the internet world is un-predictable like the electronic data theft just needs a simple action as copy and download, then the opera-

---

[7] Andrew L. Shapiro, "The Internet," Foreign Policy, Vol. II, No. 5 (1999), pp. 17-19.

tion is done. Therefore, unless there is network or IT department is keep monitoring, otherwise most people are not aware of the theft.

In the early stage, Internet attacks were simply conducted by individual hackers, but governments or other organizations have found that they have gradually been attacked by a group of hackers in recent years; national organizations have established a dedicated network attack force (here in after referred to as the "cyber army") and tailored for a specific target invasion strategy.

Originally, hackers are as Cyber criminals, and they try to invade the government service just for fun. Nobody will make the connection to hackers and Wars. Maybe the word "Hacker" is a negative one, but they just want to show off their talent in information technique. So they invade some agencies but they won't intend to destroy the system, or maybe they will only steal some data or change something information about themselves. In general, they believe that freedom is the faith, they won't from the alliance with government; actually, they are always on the rebel side to nation agency. For them, debugging or fixing security problems are their part of the life. They won't be active to attract any computer, in fact they often help information security companies to improve their product. Usually we call them "white hat" or "sneaker". However, part of "hackers" will apply their skill on some illegal things, we usually call them "black hat" or "cracker".[8] They will try to use their talent about computer to do cybercrime. For example, they will perform unauthorized remote computer break-ins via a communication networks such as the Internet. Anyway, no matter what they do (white or black hat), in the cyber space, they will follow some unwritten rules which we call

---

[8] Pekka Himanen, The Hacker Ethic And the Spirit of the Information Age (London: Martin Secker & Warburg Ltd, 2001).

"Hacktivism".[9] So the hackers are not the cyber army, they are with totally different thoughts. Recently, there is a new word about hacker" Hacktivist" , and a well-known hacker group called " Anonymous" is one part of them. Anonymous became famous for a series of well-publicized publicity stunts and distributed denial-of-service (DDoS) attacks on government, religious, and corporate websites.

But, for some national organization, they want to get the help from the sneaker, because those sneakers can get the intelligence easily. Compared with the way training by themselves, outsourcing can save more budget from government points of views, moreover, it can avoid some legal problems as well. Besides, sneaker also can get abundant reward from government and the new cyber warfare combination of intelligence（HUMINT, GEOINT, MASINT, OSINT, SIGINT）, technology, psychology. It's not only an approach about getting information, but also a way of attack.

Therefore, different with the previous behavior of virus release or entering Back Door, the cyber army has moved from passive to active intrusion targets, or through advanced persistent threat (APT)[10] attacks and social engineering, to try and break through the information security protection network from a computer to spread through the entire system. Once the information security incident occurs, all of the linked units will be immediately affected; as long as there are bugs in the information security, it will endanger the whole system. Although it's hard to solve the current operating software vulnerabilities issues by generally, information security awareness and usual security measurements are responsibilities to everyone. Due to the rise of e-commerce and social network systems dependence, governments, pri-

---

[9] Vic Hargrave, "Hacker, Hacktivist, or Cybercriminal?" Trend Micro, June 17, 2012, <http://fearlessweb.trendmicro.com/2012/hackers-and-phishing/whats-the-difference-between-a-hacker-and-a-cybercriminal/>.

[10] 邱銘彰，〈揭露網路威脅秘辛 40 分鐘搞懂 APT〉，發表於「換個腦袋作資安」資安趨勢論壇（台北：資安人，2011 年 12 月 6 日）。

vate enterprises and even private individuals have some different level of information security awareness or protection; but information security is a process, not a product. Information security cannot be resolved just with simply buying antivirus software and setting up a firewall.

In the past, cyber-attack just happened in virtual space. However, the effect of cyber-attacks does not just simply steal information; it is evolved into an invisible war with large-scale destruction, which causes human casualties as well in recent years. Specifically, NCW is the main type about modem war, and the cyber is the link from sensors, shooter, command unit, decision maker from platform-centric warfare become to a "NET".[11]Let each unit can get information at the real time to control the situation awareness. C4ISRis the key issue of the NCW. Therefore, if the node under attack or link were destroyed, it's will lead to military defeat. That's what PLA wants to study acupuncture warfare. Although in average PLA still has a gap with America Force, they try to use the" Asymmetric warfare" to confront US. Maybe PLAAF can't face USAF directly, but they can hack into US computers to get the information about America fighters.

Furthermore, in critical infrastructure protection (CIP), more than 90 percent of U.S. infrastructure such as: electricity, water, transportation and financial systems are controlled by computer system over a communication network connection.[12]

When affected by interested parties, it may cause large-scale blackouts, or public transportation system paralysis, or stock trading system paralysis, causing panic in the people as well as accidents. [13] To summarize what cov-

---

[11] Department of Defense, "Network Centric Warfare, Department of Defense Report to Congress," July 2001, pp.1-9.

[12] Richard A. Clarke and Robert K. Knake, *Cyber War* (NY: HarperCollins, 2010), pp. 70-75.

[13] 〈近 8 成關鍵基礎設施在 2010 年遭駭客入侵〉,《網路資訊》, 2011 年 4 月 7 日, <http://news.networkmagazine.com.tw/classification/security/2011/04/07/23452/> 。

ered above, cyber warfare is moving from the virtual world into real life; from digital space into real environments.

II . PLA Cyber Army Characteristics and Conduct

The PLA brought up the concepts of cyber warfare very early as in 2002, when the army General Staff of four (Electronic Countermeasures Radar Department ：電子對抗雷達部) Minister Major General Dai Qing-Min（戴清民） revealed in an internal report with a total ten type of "information warfare"(信息戰) and which was focused on "Integrated electronic network warfare"（網電一體戰）.[14]

It was declared in the report of 'Military Power of the People's Republic of China 2008 from The U.S. Department of Defense: "Strategies of PRC civilian and military have debated the nature of modern warfare. These debates draw on sources within the PLA strategic tradition and its historical experiences to provide perspective on the "revolution in military affairs," "asymmetric warfare," in the early period of PLA military campaign was to dominate the electromagnetic advantage to ensure the primary task of the battlefield victory.[15] Integrated electronic network warfare" is described as the use of electronic warfare, computer network operations, dynamic ways of killing enemy combatants, blocking the operations of enemy support, and attacking enemy network information systems（Nodes） on the battlefield. Therefore, "Integrated electronic network warfare" is seen as one of the basic forms of "integrated joint operations".

The Academy of military Sciences PLA- internal （軍內發行） compilation volume "Battlefield Network Warfare"（戰場網路戰） offers a picture of this theoretical foundation in greater detail, describing the role of

---

[14] 林勤經,〈中共網軍建設與未來發展〉,林中斌主編,《廟算台海》(台北：學生書局,2002 年),頁 439。

[15] 劉宜友,〈淺析中共網電一體戰〉,《國防雜誌》,第 26 卷,第 3 期 (2011 年),頁 121。

deterrence in the specific context of network warfare strategy and top-level policy guidance (網路戰方針). "Battlefield Network Warfare" is the culmination of a five-year collaborative research and consultation project on information warfare theory that began since 1997,and the findings represent a consensus assessment of PLA military theory on network warfare.[16] Because of the Third Taiwan Strait Crisis, It makes PLA understand the importance of information. Although PLA gets some modern weapon from Russia, they still can't evenly match with US, especially in the strategy thinking. From open resource, we can know PLA focus on establishing basic hardware system in China from1997 to 2000 .Therefore, on PLA newspaper, some authors start to use cyber Army to discuss information warfare, that's mean PLA finish the foundation about cyber warfare.[17]

In 2000, the Master of information warfare :Shen Wei Guang （沈偉光）who is the first officer to spread Information war in PLA, indicated some important theories in PLA and with his point of view, the Information war is just a form of war therefore, Information war can be applied in many kinds of the attack. Besides of being the first officer and spreading Information war in PLA, he suggested PLA with the necessity of setting up the information warfare school, or cooperation with civil school in China.[18] Nowadays in China, PLA still follows his guide with the suggestions.

Cyber warfare officially appeared in the Kosovo war for the first time in 1999. At that time, hackers invaded into NATO network systems through computer viruses which attached in the e-mail, even intended to replace the

---

[16] Joe McReynold, "Chinese Thinking on Cyber Deterrence", paper presented at the conference of "The PLA Prepares for Military Struggle in the Information Age: Changing Threats, Doctrine, and Combat Capabilities" (Taipei: CAPS-NDU-RAND 2013 International Conference on PLA Affairs, November 14-15, 2013).

[17] 張明睿，《解放軍戰略決策的辯證》（台北：黎明出版社，2004 年），頁 256-258。

[18] 沈偉光，《解密信息安全》（北京：新華出版社，2003 年），頁 119-123。

official NATO website homepage to protest against the bombing.[19] At this moment, a US stealth fighter F-117 was shot down by ground AA Gun. It's an important lesson to US about information warfare.[20]

The subsequent mistaken bombing of the Chinese Embassy in the United States was claiming by the hacker group "Honker Union of China", （中國紅客聯盟） and later they began cyber-attacks in the United States. After former Taiwanese President Lee Teng-hui（李登輝） made a special cross-strait relations statement（特殊國與國關係） in 1999, this led the CPC hackers against ROC government, universities, and commercial websites, to replace the CPC's flag and March of the Volunteerson（義勇兵進行曲）to the homepage on ROC websites. Taiwanese hackers launched a counter-attack, which raised a hacking war between the both sides .On May 1, 2002, the official websites of the White House, the State Department, and the Pentagon, were all attacked by Chinese hackers (replacing the homepages). Although the attackers from the hacking group were claiming that they were non-governmental, and "Honker Union of China（紅客聯盟）," was the group irrelevant to the PLA.[21] These internet hackers in the early 21st century mainly bring the harassment, but while people have increasing dependence on information systems for science and the form of the warfare（NCW）, the impact of cyber-attacks becomes a stronger effect.

1. E-Commerce and Information Security Vulnerabilities

Human relies on technology deeper and deeper due to the rise of e-commerce: E-commerce is borderless globalization, which is with an emphasis on intangible things, ideas, information, and the relationship that between how these three concepts are connected to come out a network of

---

[19] 東鳥，《中國輸不起的網路戰爭》（長沙：湖南人民出版社，2010 年），頁 45。

[20] 王正德，《決戰賽柏空間》（北京：軍事科學出版社，2003 年），頁 282。

[21] 劉台平，《島計畫》（台北：時英出版社，2004 年），頁 45。

electronic marketplace and society.[22] In the activities of using e-commerce, no matter the applications of bank wire transfers, ATM and credit cards, it represents that information technology network applies in human life and creates huge business opportunities, therefore, it becomes more attractive to hackers as well. There were a great quantity of personal and credit card information from bank customers got stolen due to internet banking site vulnerabilities and attacked by hackers since 2010.[23] According to the report from staff at US chain target, it was failed to stop the theft of 40 million credit card records even with an escalating series of alarms from the company's computer security systems.[24]

With the rising and developing in "Data-mining"（資料探勘）area, it is also given great attention to hackers due to Data Mining is the result of statistical applications which record buying behaviors of customers. By quantifying the results, it is expected to figure out the relation between buying behavior and products. Due to it requires a lot of consumers' personal information, such as details of purchases in the past and surveys had done before, all information would be collected and sent to a computer for running statistical analysis. If someone who is interested and seeks the access to obtain personal information through security vulnerabilities, then he can easily use the stolen information in illegal purposes. With using a false identity, fraud may also illegally conduct and deal people's information over the telephone directly, and that is a problem caused by information leak age. In other

---

[22] Jeffrey F. Rayport and Bernard J Jaworski 著，黃士銘、洪育忠譯，《電子商務》(E-Commerce)（台北：麥格羅希爾出版社，2003 年），頁 2。

[23] 〈客資外洩玉山銀遭罰 400 萬〉,《中央通訊社》, 2010 年 12 月 9 日，
<http://tw.news.yahoo.com/%E5%AE%A2%E8%B3%87%E5%A4%96%E6%B4%A9-%E7%8E%89%E5%B1%B1%E9%8A%80%E9%81%AD%E7%BD%B0400%E8%90%AC.html>。

[24] John Leyden, "Target ignored hacker alarms as crooks took 40m credit cards- claim," *The Register,* Mar14, 2014,
<http://www.theregister.co.uk/2014/03/14/target_failed_to_act_on_security_alerts/>.

words to the hackers, the database is a source of profit with selling information to other groups for use,[25] which makes terrorist organizations and
the underground economy become more rampant because of those selling
behaviors. It also represents the invisible threats in the near future of being
attacked from hackers since there are tons of personal information of customers in the banking, community websites, shopping malls and cloud services. In information age, criminal gangsters not only use the traditional vi-
olence, but also information power to create huge profits. You don't know
who use your personal information, where they are, or what they do. They
just only get your ID number, birthday, or credit card data, and they can create another you.[26]

Cloud services are based on the user interface; with the local network
connection, users can obtain remote host network services by terminal
equipment and without limitation of time and place. It is a concept evolved of
distributed computing and grid computing.[27] Cloud technology emphasizes:
multi-user, large-scale, high flexibility and self-service resources through the
information data files that are kept in cyberspace,[28] which means consumers
can access through smart phones, laptops, tablet PCs, and even desktop
computers. Any "access point" can simultaneously receive the same data
shared from a "Cloud". Cloud technology provides a high level of convenience, but it also brings a high level of risk, as hackers know that they can
attack directly the cloud database because it has the most complete files.

---

[25] Bruce Schneier 著，吳蔓玲譯，《秘密與謊言》(*Secrets & Lies*)（台北：商周出版，2001
年），頁 41。

[26] Frank W. Abagnale 著，黃維明譯，《你詐不到我》(*The Art of the Steal*)（台北：新新聞
出版社，2003 年），頁 192-198。

[27] 王平等，〈雲端運算服務之資安風險與挑戰〉，《資訊安全通訊雜誌》，第 16 卷，第 4
期（2010 年），頁 45-49。

[28] Tim Mather, Subra Kumaraswamy, and Shahed Latif 著，胡為君譯，《雲端資安與隱私
企業風險應對之道》(*Cloud Security and Privacy: An Enterprise Perspective on Risks and
Compliance*)（台北：碁峰資訊，2012 年），頁 8。

Although cloud service is convenient to users, it still contains high risk in file protecting. Because the data is too concentrated in same place, once the cloud fails or attacked by hackers, the losses will be even more difficult to estimate. Ideally, Cloud companies should have higher level of information secretly, however, "when rogues go in procession, the devil holds the cross". Not only the new attack type, but also human weakness. For example, in the military base or some government offices, it is prohibited to use USB Flash Drive and everyone follows this rule. But, if there's just one who is careless in the rule, it might get destroyed to all secretly system. While we consider having the convenience and technology advance in using of USB Charger, it provides hackers opportunities to invade the computer.[29] A small leak will sink a great ship, it's the same as the application of cloud. Even the Cloud is very convenient but it is just like putting all your eggs in one basket.

China, in addition to cyber Army, also takes advantage of business opportunities to engage in cyber warfare without restraint. While China is currently one of the world's leading producers of electronic components, many electronic factories and information companies choose China as the location to set up their host servers in Asia, and it provides a good chance to the PLA for using commercial interests as a cover. The U.S. House of Representatives Intelligence Committee announced that hat had investigated China leading telecom equipment companies Hua-wei Technologies Co. Ltd（華為）. and ZTE Corporation（中興通訊）due to suspicions of both companies are well connected to the PLA; besides, they also remind U.S. companies for not co-operating with these enterprises. It's similar between Hua-wei and PLA due to both of them have a party committee inside of the organization, and

---

[29] Angelos Stavrou and Zhaohui Wang, *Exploiting Smart-Phone USB Connectivity For Fun And Profit* (DC: blackhat, 2011), *Black Hat*, <https://media.blackhat.com/bh-dc-11/Stavrou-Wang/BlackHat_DC_2011_Stavrou_Zhaohui_USB_exploits-Slides.pdf>.

Hua-wei has a group of professional people provide services to PLA.[30] Alt-hough it did not find any evidences of spying activities from Hua-wei in the subsequent investigation, consumers should get attention to that Hua-wei's technology products are considered as easily to be hacked because of infor-mation security vulnerabilities.[31] Moreover, because of the rising economy in China, many brands are getting popular in these days as Miui （小米機）, and Chang Jiang（長江牌手機）which are famous smart phone brands from China and with good market share all over the world. Not only hardware mentioned before, but also software companies are included, that there are many Chinese software companies have their own design products as App and antivirus software, Consumers will never know if those software prod-ucts are certainly safe or not even with declarations or agreements. Once consumers use those products, they are still under the risk of information secretly. Especially, there are some illegal platforms which allows people for free to download information like (BT), we could definitely say that is the dangerous and just like to open the door for welcoming hackers.

2. Security Crises Triggered by Leaked Information

With the popularity of social networking websites and network com-munications software, information security issues have enhanced to a new level. Facebook contains a number of emerging platforms, and with Face-book comes to the public, it brings people to another layer of online world. When using Facebook and sharing information, people often reveal a lot of personal information not on purpose. With some marketing promotions of "Check In" function from Facebook, it becomes one of the most popular ap-

---

[30] 何怡蓓，〈美國指華為中興疑涉間諜活動 美大選貿易保護主義是始作俑者〉，《亞洲週刊》，第 26 卷，第 42 期（2012 年），頁 8。

[31] 康彰榮，〈未發現華為從事間諜活動〉，《中時電子報》，2012 年 10 月 19 日，

<http://tw.news.yahoo.com/%E7%99%BD%E5%AE%AE-%E6%9C%AA%E7%99%BC%E7%8F%BE%E8%8F%AF%E7%82%BA%E5%BE%9E%E4%BA%8B%E9%96%93%E8%AB%9C%E6%B4%BB%E5%8B%95-213000475.html>。

plications in the Facebook system. However, using this function which means to expose in details every movement from users to this world. The popularity of smart phones has allowed mobile internet capabilities to be more convenient. Although it represents the convenience to users who can surf the internet at all times via these high-tech products, on the other hand, it indicates that other people can track your situation through the internet surfing records. In your Facebook Friends list, it shows frequently the personal networks and user information that might be infringed by others and getting information of some daily schedule or living habits. All of these things and data will become the reference and get attacked by hackers in the future. There was an America Navy Admiral and also NATO officials who was considered to get hacked and used by spy, establishing fake Facebook account and getting other information from Facebook friend list.[32] Continuously, the spy created another fake account of this Admiral and tied to add friends of other officials in order to get more information of intelligence until got exposed by other people.[33] In this case, hackers didn't use any high technology; however, they got the intelligence. Therefore, leaks of information resulted in the fraud issues: the bigger problem is the use of false identity to defraud the trust from others. More and more hackers exploit vulnerabilities in e-commerce to engage illegal financial transactions and thus destroy the world financial order. Generally speaking, cybercrime is a major threat and could jeopardize national security. In addition to external hackers, there are many breaches of data due to neglect, and customer-related data were destroyed which were resulted from time and cost considerations as job transfer or office migration. Therefore, for interested par-

---

[32] 〈冒名北約統帥/中國網諜臉書竊密〉,《自由電子報》,2012 年 03 月 12 ,
      <http://www.libertytimes.com.tw/ 2012/new/mar/12/today-int1.htm>。

[33] Emil Protalinski, "Chinese spies use fake Facebook profile to friend NATO officials,"
      ZDNet, March 11, 2012,
      <http://www.zdnet.com/blog/facebook/chinese-spies-used-fake-facebook-profile-to-friend-nato-officials/10389>.

ties, it is easy to obtain the relevant data. With the current extent of information technology, using hard disks and electronic database is more and more common than before. . For hackers, no doubt, this makes the access to information become more convenient, which means they can simply copy and transfer large amounts of data in a short time. The private sectors especially do not have tight protection systems like the government or the military. Insurance companies, airlines, banks, and various shops give out membership cards for business needs, and then collect a large amount of personal data from customers. But it only needs a hard drive as USB flash drive and all information could be copied to be a great loss in a short time.[34]

This is currently one of the serious reasons behind data leakage. Although many companies have firewall or use other system to protect their data, they often outsource from other companies. If there are bugs in the system, it will be rare to find out by client side even they will update the software. In April, 2014, The "Heartbleed Bug" CVE-2014-0160 is a particularly nasty bug about OpenSSL It allows an attacker to read up to 64KB of memory.[35] It means many Digital Products will be invaded, including smart phone in Google Android system, and many servers companies （Cisco Systems、Juniper Networks）are also in the high risk list.

Personal information leakage is a serious problem, but most people don't still realize this issue with the same level of importance as national security. Personal data leakage allows hackers to understand your career, tastes, and relationships with friends. Armed with this information, a network hacker could attack specific important targets by forwarding mail har-

---

[34] 〈買二手硬碟赫見銀行往來資料〉,《聯合新聞網》,2012 年 2 月 29 日, <http://udn.com/news/LIFE/LIF1/6930184.shtml>。

[35] David Grant, "The Bleeding Hearts Club: Heartbleed Recovery for System Administrators," *Electronic frontier Foundation*, April 10, 2014, <https://www.eff.org/deeplinks/2014/04/bleeding-hearts-club-heartbleed-recovery-system-administrators>.

boring malware.[36] In the past, it could be protected by antivirus software of the attack with Trojans attached due to there were too many samples using the same methods and it brought the attention to antivirus software companies.[37] However, in the plan of an organized hacker group or even national cyber army, they have precise strategies which can make victims get targeted in and lose their vigilances or even pass their personal information like password directly because of some human weaknesses.. When the victim opens an interesting file, it will start a Trojan program or open a backdoor intrusion system, thereby breaking through the firewall and network information security protection, and obtaining sensitive information. Particular cyber army groups organized by hostile countries could steal information of operational electronic parameters, such as research data of military arms, equipment or personnel's background of intelligence which would cause seriously harm to national security. The PLA Cyber Army repeatedly invaded BAE system of the British defense industry and tried to steal the relevant data of the cooperation development with the U.S. for the F-35 stealth fighter, which will affect the future of the F-35's survivability on the battlefield,[38] as well as using the data to assist in the development of their own advanced fighter（J-20, J-31）. Therefore, through computer intrusion, use of personal data theft via information security systems, provides a new means for company and state espionage.

New types of cyber-attacks are entirely different from traditional viruses and Trojans, which implanted into computer that focused on the intrusion through the network to achieve the purpose of the invasion and paralyze other information systems. At present, the sticking point in tactics and the use of a

---

[36] Martin C. Libicki, *Cyberdeterrence and cyberwar* (CA: Rand Company: 2009).

[37] Dorothy E. Denning 著，戴清民、吳漢平譯，《信息戰與信息安全》*(Information Warfare and Security)*（北京：電子工業出版社，2003 年），頁 388-386。

[38]〈中國駭客入侵 BAE 竊取 F35 機密〉,《自由電子報》, 2012 年 03 月 13 日，<http://www.libertytimes.com.tw/2012/new/mar/13/today-int5.htm>。

personal information acquisition is to break the defense. The appearance of social engineering, APT attacks and watering hole attacks are the new methods of attack.

Ⅲ. PLA Cyber Army New Attack Types

1. Cyber Army and APT Attacks

Advanced persistent threat (APT) attacks emerged in 2005 as a hacker attack tactic. It uses the methods of human nature and social engineering to gain victims' trust; and it is currently the way that PLA cyber army uses frequently. .APT has the three main features which are persistent, latent, and high aggressiveness attack.[39] APT attacks are an organized plan of espionage, and are different from information security incidents in the past as previous hackers used the opportunity to show off their abilities. The main purpose of APT is harassment and mischief through replacement webpages or use distributed denial of service (DDOS) attacks to try paralyzing the host server, in order to demonstrate their superior technology. In the past, hackers looked for the most vulnerable place to attack, but in APT attacks, a professional team needs to gather information for a long time before identifying vulnerabilities, data analysis, environmental testing and data acquisition. This requires a lot of capital investment for completion which is not generally suitable for ordinary hackers. Therefore, the target value selection is emphasized in the choice of attack order.[40] From the targets of government to Individuals, APT attacks are planned, organized and targeted by malicious network activity, especially for the public sector penetration. The term APT first appeared in the U.S. Air Force internet security report in 2006, with a group of digital soldiers who were organized by government and well trained. . The important goals of social circles, suppliers, and customer lists

---

[39] 李青山，〈APT 發展趨勢研究漫談〉,《信息安全與通信保密》, 2012 年 07 期（2012 年），頁 19-21。

[40] David Dewalt, *Definitive Guide* (MD: Cyber Edge Group LLC, 2011), pp. 17-27.

were expanded to monitor the effectiveness of this attack. The new APT technology does not focus on attack, as its feature is a potential large collection of information over the long-term, so the process is slow and low.

Although it does not technically break through with traditional Rootkit technology,[41] APT attack prevention cannot simply rely on antivirus software or a firewall. The target must be psychologically prepared for a long-term confrontation. In addition to a long incubation period and organization and planning, more 'intelligent' APT attacks will continue to try new ways for computer intrusion.

APT attacks are not the same as general e-mail viruses, and antivirus software cannot easily intercept this treat. A feature of this type of cyber-attack is that victim is difficult to realize that information has been leaked, as digitized data can be easily copied for transmission.[42] It looks like an iceberg, when information security department finds out the problem; it has occurred huge data losses already. APT attacks are generally using the way to disguise and attached as a PDF file or WORD file, and then send out with an interesting title to the target, that attract victims in opening the Rootkit file which then enters into the victim's computer.[43] With the vulnerabilities of company information security, the malicious programs could be easily broke into the defense.

In 2006, the PLA Cyber Army took advantage of a well-known defense legislator's account and sent out the e-mail military media reporters. Many reporters thought mistakenly the e-mail was a press release or relevant information, and infected unwittingly their own computers after opening the

---

[41] Greg Hoglund and James Butler, *Rootkits: Subverting the Windows Kernel* (MA: Pearson Education, 2006).

[42] 徐偉，〈APT 攻擊-狼來了及應對措施思考〉，《信息安全與通信保密》，2012 年 07 期（2012 年），頁 17。

[43] 黃耀文，〈福爾摩斯兄弟性格差異：主動式 APT 之追蹤與偵測技術分享〉，發表於「2014 亞太資訊安全」論壇（台北：資安人，2014 年 3 月 20 號）。

attached file.[44] Although the sender address and name may have subtle discrepancies, most people are not diligent enough to find the problems. In addition to legislators, the PLA Cyber Army collects systematically information on relevant national defense, foreign affairs, and political academic issues of relevant parties. At the appropriate time the relevant documents are used for the purpose of breaking through information security protection. In August 2012, the PLA Cyber Army even posted faked Defense Department documents directly on university internet websites, which tried to spread false news and Trojans in the attached archive. These practices take advantage of people's lack of observation or carelessness while downloading files directly, so that the computer virus or Trojan is successful in direct invasion.[45] The White House also received a "phishing" letter issued by PLA cyber army while users got attracted and opened the mail with attached files; it would invaded the computer automatically and started to track the users.[46]

The APT attack and management processes can be learned as the cycle of PDCA (Plan, Do, Check, Act). In the 'Plan' level, the hacker group is trying to lock on the target and collect information. The 'Do' level is the execution stage is the invasion of network systems by the hackers to create an attack platform. In the third 'Check' step, the senior command of the hacker group intelligence unit will analyze and interpret the resulting information. In the final 'Act' stage, hackers will attempt to use the victim's address books and social network to increase the validity of continued attacks.[47]This

---

[44] 黃敬平，〈記者收毒郵　疑中駭客搞鬼〉，《蘋果日報》，2006 年 4 月 14 日，
<http://www.appledaily.com.tw/appledaily/article/headline/20060410/2528327/>。

[45] 吳明杰，〈假公文臺軍漏餡　疑中共網軍散佈〉，《中時電子報》，2012 年 8 月 20 日，
<http://tw.news.yahoo.com/%E5%81%87%E5%85%AC%E6%96%87%E8%87%BA%E8
%BB%8D%E9%9C%B2%E9%A4%A1-%E7%96%91%E4%B8%AD%E5%85%B1%E7
%B6%B2%E8%BB%8D%E6%95%A3%E6%92%AD-213000848.html>。

[46] 劉項，〈中國駭客入侵美國網絡〉，《亞洲週刊》，第 26 卷，第 41 期（2012 年），頁 9。

[47] Roamer，〈APT 教戰手冊〉，發表於「擋住資安骨牌效益」研討會（台北：資安人，
2011 年 8 月 11 日）。

is the typical cycle of APT attacks. With more and more similar emerging cases, the military conducts important exercises every year（ex: 漢光演習），as the number of e-mail attacks increases. But APT attack still needs cyber army to take the initiative; recently, they change the method. That's what we call "watering Hole" attack（水坑式攻擊）.[48] By this way, cyber army doesn't need to invade targets; all they have to do is like the lions setting the trap near the water. So they will send a link to your mailbox which with the interesting websites. When you check it, cyber army will set a "drive-by"（偷渡），and combine the 0-day attack, then your computer would be under attack.

Many experts consider that these attacks are from the PLA Cyber Army. Due to the complex nature of cross-strait relations, Taiwan has become the test platform of PLA Cyber Army tactics. Cyber Army troops also use a network of number Taiwanese host servers when jumping between versions. After the invasion by malicious computer programs, the computer is controlled by hackers, and the host attacker sends junk information to the host server.

When zombie computers are under control of Botnet, the exploited victim can only be discovered by tracking the relevant legal IP address. The antivirus software company Symantec reported that there are 7% of zombie computers of the world in Taiwan and Taipei is the city with the most zombie computers for 5% of the world, These computers will automatically connected and controlled by the hackers center and wait for a command, which means they are always ready to attack others.[49]

In recent years, the PLA Cyber Army launched several APT attacks that considered as threats to all countries. Google's 'Operation Aurora' （極光行

---

[48] 〈2013 上半年安全報告：水坑式攻擊成為重要安全威脅〉,《資安人科技網》, 2013 年 09 月 03 日,< https://www.informationsecurity.com.tw/article/article_detail.aspx?aid=7648 > 。

[49] 王平等,〈僵屍病毒解藥與監控系統之研發〉,《昆山科技大學學報》, 第 9 期,（2012 年 6 月）, 頁 37-47。

動）[50]that report in 2010, and the McAfee security company published the 'Night Dragon' （夜龍行動）in the 2011[51]. In the report of PLA cyber army action, there was a list with 10 APT attacks targets which were mainly energy companies. They used external host web servers to start the attack, then using SQL to put injection attacks, and afterwards their repeated to use as a springboard to achieve further attacks in the intranet. The repeated to use those remote access tools (RAT) to return a lot of important information form WORD, PPT, and PDF files. In August of 2011, an information company released a security report, which stated that there is one "Nation" guided the actions of network intrusion, which with code-named "operation: Shady RAT"（暗鼠行動）.[52] This name was coincides with "Remote Access Tool" (RAT), while "Rat" also represents a creature that is lurking in darkness, and trying to destroy or steal valuable items all the time. This coincides with current hacking and intrusion methods. It's like the "mole" in intelligence term.

The whole operation lasted five years. The targets were covered in the United States, Taiwan, South Korea, India and major international organizations, such as the United Nations, ASEAN, major news media companies, many Anglo-American defense contractors, and many key infrastructure-related manufacturers. As well as the United Nations, the Republic of China government, defense contractors, a number of international companies and seventy-two organizations have been hacked. According to the analysis of security information, all these hacks originate from the China. In addition to

---

[50] Xecure-Lab 研究團隊，〈三起 APT 事件攻擊手法解析〉，發表於「2012 資安趨勢專刊 -換個腦袋作資安」（台北：資安人，2011 年 12 月 6 日），13 頁。

[51] Professional Services and McAfee Lab, "Global Energy Cyberattacks: Night Dragon," *McAfee*, February 10, 2011, <http://www.mcafee.com/ca/resources/white-papers/wp-global-energy-cyberattacks-night-dragon.pdf.>.

[52] Scott Jasper 著，董光復譯，〈美國與中共的網路之爭〉 *(Are US and Chinese Cyber Intrusions So Different?)* ，《國防譯粹》，第 40 卷，第 12 期（2012 年），頁 77-79。

attempting to invade computers, hackers try to directly invade Gmail accounts to view user's e-mails, and sometimes create a similar account to palm of your friends.

2. The Cyber Army and Social Engineering

Social engineering is a superb technique that does not require invasive methods. The victim is deceived when mentality off guard, by dint of conversation skills and forged documents, the fake identities would obtain secrets from legitimate users, such as account passwords or relevant information. While most information security staff put emphasis on software and hardware technologies, they ignore that human nature is the easiest point of attack to neglect.[53]Social engineering is now mostly in the usage of e-mail scams, often as junk e-mail advertising adult websites or shopping sites as a cover. But when combining with APT attacks, social engineering analyzes the habits of the target, as well as: habits in news reading, types of particular interest articles which , the most frequent receiver lists of e-mail, the relationship between the e-mail sender and receiver, and determining a certain time to send an e-mail which is containing malware to deceive other victims.[54] Social engineering uses a variety of basic human emotions, such as intimidation, playfulness, compassion, flattery, authority, pressure, vanity and so on to find ways defrauding the trust of the target. People are the most vulnerable part of information security, as human nature cannot be repaired or received software upgrades. [55]

Although information security systems costs a lot of budget, the destructive poser of social engineering will totally destroy the system in a short

[53] Kevin D. Mitnick and William L. Simon 著,子玉譯,《駭客大騙局》(The Art of Deception: Controlling the Human Element of Security)（台北：藍鯨出版社,2003 年 9 月）,頁 14。

[54] Johnny Long, No Tech Hacking: A Guide to Social Engineering, Dumpster Diving, and Shoulder Surfing (MA: Syngress publishing, 2008), pp. 110-117.

[55] Cyrus Peikari and Anton Cbuvakin 著,陳建勳譯,《防駭戰士》(Security Warrior)（台北：美商歐萊禮股份有限公司,2007 年）,215 頁。

time., The company HB-Gary which is international famous company and with the main customer as U.S government and Pentagon. They play an important role to provide information services which are covering identities of hackers, penetration performance testing, and information security protection. But this company was invaded by hackers who disclosed a lot of information, to the extent that the company was forced to quit the RSA2011 information Security Annual conference in February of 2011.[56] IT The most important key is that hackers posing as one the company's top executives sent an e-mail to the IT department manager, claiming that during their meeting in other country, there was a desperate need to change the password and settings. The IT department manager made the changes, with predictable results.[57] Social engineering has become a viable method to bypass password protection, computer security and defense, network security and defense, and all technical security measures to directly attack the most vulnerable part of the system, humanity. No amount of firewalls can stop the power of an e-mail.

Another famous case is the RSA subsidiary information security division of the EMC Corporation, which developed the one time password SecureID system. In March of 2011, RSA publicly declared that their technique had been hacked by using APT attacks, and the information of product certification has been leaked. So, about APT attacks, it's beyond cyberspace, and it's a war. The first thing was choose the Target. For example, PLA wants to get the data about F-35 from Lockheed Martine, secondly, PLA cyber army will try to find the system information about Lockheed Martine, therefore they focus on the RSA Security. Because RSA is a famous information company, and not only Lockheed Martine but also Northrop-Grumman are

---

[56] 胡曉荷、白浩、周雪,〈RSA 大會-信息安全界的不老傳說〉,《信息安全與通信保密》, 2012 年 04 期(2012 年),頁 9-13。

[57] Nate Anderson, "Anonymous vs. HBGary: the aftermath," *arstechnica*, February 25, 2011, <http://arstechnica.com/tech-policy/news/2011/02/anonymous-vs-hbgary-the-aftermath.ars>.

both clients of RSA. That's the main reason why they choose RSA. RSA SecurID is key stone of technology about identity. However, Cyber army finds the employee in RSA by brewing blog. Then cyber army starts their plan.

The origin of the RSA security information being compromised is well known, as two e-mails that could start the whole episode. Using an Excel file as a cover, with that file which was implanted with Adobe's "zero-day" (零時差攻擊) [58] Flash flaw (CVE-2011-0609) and then when it be opened, it could be operated by a remote control tool. This Allowed direct system server invasion as well as invasion into other system server at the same time. Many important companies and arms dealers are using RSA's system, so the data leakage events lead indirectly to the hacking of the important defense industry company Lockheed Martin,[59]where the large amounts of data were stolen. This was due to the combination of APT attacks and social engineering

This was almost exactly the same process of attack that Google encountered in Operation Aurora. Google Inc. indicated in their report, which was published in January of 2010. It was the PLA with carefully planned and targeted on this attack, moreover, PLA was considered that linked with specialized electronic hacking training schools where located in Shanghai's Jiao Tong University （上海交通大學）and Shandong's Lanxiang high School （山東藍翔高校）.[60] This report also pays attention to the use of the cyber army within the military forces, and the use of national resources and a rig-

---

[58] A zero day vulnerability refers to a hole in software that is unknown to the vendor. This security hole is then exploited by hackers before the vendor becomes aware and hurries to fix it—this exploit is called a zero day attack.

[59] "The U.S. Arms Manufacturer Lockheed Martin Hacked Websites Affects Pentagon," CNYES, May 30, 2011,
<http://tw.money.yahoo.com/news_article/adbf/d_a_110530_2_2lws9>.

[60] 管淑平，〈Google:駭客技術來自中國〉,《自由時報》，2010 年 2 月 23 日，版 A8。

orous academic curriculum to train the cyber soldiers.[61] As well as "information security professionals" (信息安全人才) training in police departments, military units have also established related departments to train network talent.[62] A selection of the cyber army elite consists of the "Blue team"
division, which specifically evaluates network systems to identify vulnerabilities that could harm military readiness capabilities. There is also more
training of assets outside official channels, in order to recruit a large number
of "Patriotic hackers"(愛國黑客) that possess a certain degree of computer
literacy. These non-government personnel are willing to launch cyber-attacks
against an adversary. Although the government can hire such professionals to
serve as full-time staffs, however, China uses these patriotic hackers for independent network operations. In this way it can be argued that the criminal
behavior is being conducted by private individuals, it's nothing related to
PLA governmental responsibility. This kind of action is encouraged by PLA
approximately 250 independent hacker groups.[63] Due to the popularity of
information technology and network facilities, such "people's militia"(民兵)
has created a new era of threats to other countries. "New People's war"(新人
民戰爭) is sufficient to achieve the purpose of unity of the military and
civilians,[64] as the CPC government uses a direct appeal to join the ranks of
patriotic hackers for conducting network operations.[65] The government is
deliberately cultivating and training new personnel in order to expand its
combat effectiveness in the internet war. Those network soldiers of govern-

---

[61] John Bumgarner 著，周敦彥譯，〈確保網路安全：重新思考新防禦時代之軍事準則〉
(Securing the Cyber Sphere: Rethinking Military Doctrine for a New Defense Era)，《國防
譯粹》，第 38 卷，第 4 期（2012 年 4 月），頁 43。

[62] 趙俊閎、吳曉平、秦豔琳，〈信息安全人才培養體系研究〉，《北京電子科技學院學報》，
第 14 卷，第 1 期（2006 年），頁 27-31。

[63] Richard Weitz 著，高一中譯，〈網路作戰威脅日增〉(Firewall: The Growing Threat of
Cyber Warfar)，《國防譯粹》，第 37 卷，第 5 期（2010 年月），頁 35。

[64] 楊念祖，《決戰時刻》（台北：時英出版社，2007 年 2 月），頁 131。

[65] 張召忠，《網路戰爭》（北京：解放軍文藝出版社，2001 年 2 月），頁 62-65。

ment organization deliberate collect personal data and network usage information to target and analyzing the dependence on the system and the possible impact of the military, economic, and political after destruction. After the
official training and with relatively high-quality information equipment, the
use of a "saturation attack"（飽和攻擊） approach is tested to better understand national cyber army's combat effectiveness. And in the PLA,
"Three Warfare" as psychological, media, and legal operations,（心理戰、輿
論戰、法律戰）[66] which are their important strategies. How to use information power to improve "Three warfare"？ That's a new thinking way to
PLA. On the contrary, it's a big challenge for us to face" New Three warfare"
that combine the cyber warfare.

Ⅳ. Conclusion: Taiwan Information Security

After World War I, the German General Ludendorff Erich Ludendorff
has indicated that: in modern warfare, it is difficult to distinguish between
the front and rear action, because the people and the army are one, war has
no distinction.[67] In the concept of information security, not only defense
needs to mobilize completely establishing a system, more important is that it
requires a degree of integration from the nation as well. The establishment
requires the government, the army and interaction and cooperation with the
private sector to integrate of civilian and military bodies that jointly safeguard the cyber space. Cyber warfare is an invisible war, with no distinction
between peacetime and war time. In cyberspace, each computer network has
a possibility to be the cyber war battlefield. Especially in recent years, the
popularity of smart phones means that the internet does not require a fixed

---

[66] Timothy A. Walton, "China's Three Warfares," *Delex Systems, INC consulting Studes and Analysis,* January18, 2012, <http://www.delex.com/data/files/Three%20Warfares.pdf>.

[67] 鈕先鍾，《西方戰略思想史》（台北：麥田出版社，1999 年 12 月），頁 433。

location for operation. [68] Also with the advancement of technology, that the hacker has an increased success rate of its invasion. Every internet user has the possibility to become a new network soldier .If the PLA mobilizes its Cyber Army as well as its internet users, it can immediately launch an offensive on the network. Through the recruitment and professional training of its Cyber Army, its organization and trained combat power are extremely important. In Guangzhou's Military Region（廣州軍區）, PLA also sets up China Unicom, military units; combat divisions and the training brigade base of the "Cyber Blue team" （網路藍軍）to strengthen its cyber warfare abilities. [69]

In the report from the information security film Mandiant and Crowdstrike stated that PLA Unit 61398[70]、61486[71] is PLA Cyber Army in Shanghai （上海）, And he point out PLA Cyber Army was under PLA General Staff Department（GSD 總參謀部）. It's responsible for offensive and defensive intelligence to gather efforts directly at information and communication systems from enemy forces and political targets. And the Information Warfare Intelligence Research Center in the Nanjing Military Region （南京軍區）, where forces on Taiwan.[72] However, cyber space is a new attack type. It's beyond geographic, as long as there is internet, cyber army can act their attack everywhere. As we know, intelligence is the key points of APT attack; cyber army needs to understand the target and his personal information. In PLA, these HUMINT was controlled by second department in

---

[68] Kris E. Barcomb 著，高一中譯，〈從海權到網權：以史為鑑勾勒未來的戰略〉 *(From Sea Power to Cyber Power: Learning from the Past to Craft a Strategy for the Future)*，《國防譯粹》，第 41 卷，第 2 期（2014 年 2），頁 33-37。

[69] 羅印沖，〈中共建網路藍軍 演練四面出擊〉，《聯合報》，2011 年 5 月 27 日，版 A17。

[70] 葉茂之、劉子威，《中國國安委》（香港：領袖出版社，2014 年 1 月），頁 412-417。

[71] 中央社，〈美媒：第二隻大陸駭客部隊曝光〉，《中時電子報》，2014 年 06 月 10 日，<http://www.chinatimes.com/realtimenews/20140610004640-260409 >。

[72] New Frontier Foundation Defense Policy Advisory Committee, *China's Military Threats against Taiwan in 2025* (Taipei: New Frontier Foundation, 2014), p. 63.

GSD（總參二部）, and the Signet was third department in GSD（總參三部）.Cyber army has to get the help from them, so they will work near them. Maybe we can say PLA cyber army has a base in Nanjing or other RM, but their main command center still in China capitals: Beijing（北京）

As long as information is stored in a computer with internet capability, the data has a chance of being stolen. Even though we can use Physical Isolation Network to protect data, but because of the need to work, confidential data needs to be copied to a personal computer first, and that's where it can be taken. [73] Any tool that can access digital information and the internet that could have Trojans implanted inside. Social engineering could be used to take important relationship network details from people who work in Intelligence, or relative of officials, teachers, and friends in order to try and break through into critical information or gain passwords. Although the relationships between people have increased day by day with the spread of electronic information, through MSN, Facebook and other social networking sites, the distance between people is getting shorter and shorter, as they are all part of the same network transmission exchange. Social engineering takes advantage of this network interaction through forwarded messages and downloadable internet resources, which also provides a shortcut for hackers. Even though guarded by an impregnable security system, one person's mistake with lack of training could lead to a national security threat. If new weapons development information was leaked through cyber-attacks, or if hostile countries obtained the relevant data on the F-35,[74] it might endanger its viability in the future battlefield, leading the United States lose its air supremacy.

---

[73] 李志德，〈公務家辦 對岸網軍等者撈〉，《聯合報》，2007 年 10 月 14 日，版 A1。

[74] Richard A. Clarke and Robert K. Knake, *Cyber War* (NY: HarperCollins, 2010), pp. 233-235.

Information security is a process not a product. Cyber warfare is an invisible war, although defender is harder than offensive, we can use the defense in depth to take more space. In Taiwan Office of Information and Communication Security（OICS） of Executive Yuan is the dominant about Information security, but actually this office is just responsible for Information security. However, critical infrastructure protection is the duty of Office of Homeland Security（OHS）. This policy will confuse national information security. In this time, CI will be the first target of cyber army. As Iran's nuclear power plant, U.S. and Israel cyber army use the "Stuxnet Worm" attack Siemens WinCCS7 SCADA system, which was used to control the centrifuge in plant.

In this case, cyber army didn't use self-replicating way to spare, because it will be exposed. They use the 0-day exploit MS10_406.LNK vulnerability; USB flash desk and Windows Printer Spooler to pack pass their information security system. And they use Rootkit with "Fake" Digital Signature hidden in system to destroy centrifuge.

In Taiwan, not only nuclear plant but also CI, if their system were under attacked, that would become a disaster. In the warfare, we can understand the Joint operations is necessary, but, we don't have an integrated policy about Taiwan information security. OICS and OHS don't still enough to conform it, they focus on defense and formulating regulations. If we want to have an offensive and defensive information force, we have to integrate government, military force, and civil companies. There are many powerful hackers in Taiwan, TDOH(資安社群The Declaration of hacker, TDOH) and HITCON （台灣駭客年會） is also famous in the international arena; some of them venture the company about information security. And M&A with foreign company, many anti-APT group are on the list.[75] So how to take the civil

---

[75] 謝易宣，〈資安輸出國際 四管齊下〉，《經濟日報》，2014 年 3 月 24 日，
<http://udn.com/NEWS/STOCK/STO2/8567084.shtml>。

industrial advantages to improve Information security in Taiwan is the way what we have to do.

In the past, the emphasis was on "It is everyone's responsibility to conceal national secrets and keep alert of hostile spies."(防諜保密、人人有責) However, today's internet world we need to promote constantly drilled vigilance to achieve the desired effect. The better way is that do not to open unsolicited e-mail, or re-determine the correct communication channels when on the telephone with someone that you haven't confirmed the identity of. When surfing on community websites, it's too dangerous to reveal much personal information including "sign in" of Facebook. Carefully select friends on your network, avoiding interested parties of fraud, or those who do not protect their own privacy. More important thing is, avoid PLA Cyber Army social engineering references. In particular, military and national security issues which are of keen interest to military fans. While surfing on internet, enthusiastic writing is more likely to be exploited by the use of the "phishing pull" （魚叉釣出）approach to absorb military fans.[76] But it's not necessary to forbid totally the use of social networking websites for relevant personnel. After all, it took 38 years to accumulate nearly 50 million users for radio, 13 years for television, the internet took four years, the iPod took three years as well, however, Facebook has increased about two hundred million users less than in one year. So imposing various restrictions and penalties to prevent people entering social networking sites is unrealistic, but it still needs appropriate regulations to help in avoiding over-correction.[77] The U.S. military announced a guidebook of social networking sites in 2011, which hopes to strengthen the awareness of information security officers.[78]

---

[76] 閻東平，《正在進行的諜戰》（香港：明鏡出版社，2009 年），頁 546-550。

[77] James G. Stavrids and Elton C. Parker 著，高一中譯，〈航向網路之海〉（Sailing the Cyber Sea），《國防譯粹》，第 39 卷，第 8 期（2012 年），頁 6。

[78] Office of the Chief of Public Affairs, US Army Social Media Handbook, January, 2011, <http://www.slideshare.net/USArmySocialMedia/social-media-handbook-2013> .

To users of both sides could have a "code" to establish and confirm correct communication between each other like use with a password. E-mail exchanges habits of people in communication. During communications, it could also be used to determine the identity of both parties. Password should be avoided with a simple a combination of numbers and letters, as simple traditional passwords are easier to crack with auxiliary information technology. Through the assistance of cultural traditions and special dialects, we can create a unique password system which is also feasible method. In W.W II, the U.S. military used the Pacific region Navajo language as communication passwords, with names of birds instead of each fighter model. The U.S. gained considerable harvests in this area, and at the end of the war that Japan's Intelligence Chief also acknowledged that they could not decipher U.S. military's Navajo password system.[79] A similar situation also appeared in the 2012 China blockade on the Internet. After the February of 2012, Wang Li-jun （王立軍）incident, the China immediately blocked internet discussion of the incident, but the Chinese used homophonic written characters (王立軍become 王捕頭、周永康become 康師傅) and pop culture references to avoid network control.[80]

Information security protection must cover whole system from top to bottom in order to avoid threats. Development of cryptography can certainly encrypt messages, but technology always comes from human nature; the disintegration of the defense system is often from a weakness of human nature. Therefore, a complete introduction of information education is needed to promote investment in information. Internet is no geographical restrictions as national territory vanishes in cyberspace. Therefore, international cooper-

---

[79] Simon Singh 著，劉燕芬譯，《碼書:解碼與編碼的戰爭》(*The Code Book: The Science of Secrecy from Ancient Egypt to Quantum Cryptography*)（台北：商務出版社，2000 年），頁 231。

[80] 李永峰，〈中國當局限制網民討論重慶事件，民間智慧催生各類暗語與代碼〉，《亞洲週刊》，2012 年 4 月 8 日，頁 8。

ation and exchange of intelligence are needed to ensure network security. The future wars may firstly occur on the internet or proceed via domains. In addition, network attacks may also be combined with real military fighting, where the aim is to cut off logistical support system for relevant services from the real combat situation. In addition, much critical infrastructure is maintained and operated by private companies, but no matter whether it is civil society or government military, all have important responsibilities to maintain daily operations. Once the country suffers a series of coordinated attacks, it is bound to be caught off guard, and find itself in difficulty to withstand the attack. In addition to stealing important information, the enemy could possibly implant false information or spread wrong message to undermine the integrity of the total information, which in order to interfere with or mislead judgment. This is likely to cause confusion or even panic amongst the people and society.

To control is to foresee. Waiting and seeing are the worst things.[81] IT Internet totally changes our life, and it's become the real Digital Nervous System（DNS）[82], Although it brings the convenient life, risks comes along with as well. It's impossible to avoid those risks completely, but we can use the strategy to control them. As the lyrics says: Let the sky falls, when it crumbles. We will stand tall, or face it all together. Information secretly is just the same way of thinking.

---

[81] Andre Beaufre 著，鈕先鍾譯，《戰略緒論》(*Introduction to Strategy*)（台北：麥田出版社，1996 年），頁 192。

[82] Bill Gates, *Busness@The Speed of Thought: Using a Digital Nervous System* (NY：Warner Book Inc, 1999).

China's Military Rise and East Asian States: A Regional Security *Tour d'Horizon*

Taeho Kim

(Director, Hallym Institute for Taiwan Studies (HITS); Director, Center for Contemporary China Studies, Hallym University of Graduate Studies, Seoul, Korea)

ABSTRACT

While China's "military rise" is an issue of growing importance to regional security, it is worthwhile to note that it is not China's military modernization per se, but its ability to project and sustain power along and beyond its borders—in particular, the possibility to resolve forcefully its outstanding maritime disputes and various contingencies in the region.

This essay argues that China's "anti-access capability"—a U.S.-coined term originally developed for a Taiwan crisis—is equally applicable to other major regional cases such as the Spratly disputes and a North Korean contingency. Furthermore, notwithstanding China's continuous efforts to develop and deploy various types and classes of weapons/platforms, it is the Russian systems and technologies that are most capable and thus likely assigned to the highest mission-critical areas.

In assessing China's current and likely future military capabilities as well as their implications for the region, it is necessary to take note of the following:

1. There exists asymmetry of military capability between China and its weaker neighbors. While the PLAN is weak in several important aspects, many of its neighbors' navies are weaker still.

2. Some have argued that China's foreign policy behavior apparently became more "assertive" in 2009-2013, but it is wiser to keep in mind that China has always been assertive and aggressive when it comes to what China defines as "sovereignty and territorial issues" as well as its newest "core interest."

3. On the South China Sea disputes it is the function of U.S. presence in the theater—in the form of overseas bases and the freedom of navigation—and the PLA's own limitations to project and sustain power for an extended period of time that have largely prevented armed conflict from occurring for the past decade or so.

4. While Taiwan remains the *idée fixe* of China's diplomacy and military, it is and will be a tough nut to crack. China's recent creeping attempts for economic integration with Taiwan should be seen in this context.

5. China and Japan, two regional heavyweights and traditional rivals, will likely have a bilateral relationship that is replete with difficulties and tension. China's announcement of the ADIZ in November 2013 is only the latest manifestation of this deeper and difficult relationship.

6. For Korean security it is imperative to take into account the geostrategic and historical factors. On top of the existing military threats from North Korea, the ROK should be able to employ a) a hedging strategy, b) "limited defense sufficiency" strategy, and c) rock-solid relations with the United States.

Key Words
China's "military rise," "anti-access" strategy, Russian weapons and military technologies, the South China Sea, the Taiwan Strait, the Korean Peninsula

It is trite yet true to note that the rise of China is a dawning strategic reality in Asia and beyond. China's economy, for instance, has grown an average 9.8 percent annually for 30 years (1980-2009),[1] and its foreign reserve now exceeds US3.3 trillion dollars. Since 2010 it has replaced Japan as the world's second largest economy. For many economists China's rapid economic growth is indeed music to their ears, as they not only expect benefits from various economic transactions with China but such major economic blocs as the EU and America have remained stagnant or in doldrums.

It is thus not surprising to see that China's economic weight has funneled funds into its diplomatic clout and military might, even if China's continued economic and military growth is a double-edged sword in the eyes of its neighbors. While China's economic vibrancy has largely sustained the region's prosperity and vitality, its military modernization as well as its deliberate opacity in military issues—in tandem with its recent assertive foreign-policy behavior—has created a level of apprehension among many security analysts. Notwithstanding the official protestations by China's senior officials that it pursues "peaceful development" and other benign goals only, the "rise of China" is often viewed as foretelling a transformation in the regional balance of power.

Despite its considerable scope and complexity, on the other hand, the China debate has essentially revolved around two sets of questions: One is the various yet uncertain implications of a rising China, and the other how best other countries can cope with it, individually or collectively. There also exists a substantial body of literature focusing mostly on one of the multifaceted dimensions of a rising China phenomenon, such as economy, foreign policy, and security. Few, if any, have attempted to distill the region-wideconsequences of China's rise.

---

[1] China's economic growth rate has ever since shown a declining trend: 10.4 percent in 2010, 9.3 percent in 2011, 7.8 percent in 2012, and 7.6 percent in 2013.

This essay joins the China debate by illuminating the security dimension of China's rise at the regional and country level. The first section addresses the meanings of China's "military rise" in U.S.-led East Asian security and its implications for their bilateral "competitive interdependence" ties as well as for East Asian security. The next section explores its possible manifestations for East Asia's sub-regions with particular contingencies in mind—in the counter-clockwise order from Southeast Asia (i.e., the South China Sea) to the Taiwan Strait (Taiwan) to Northeast Asia (the East China Sea and the Korean Peninsula). Finally, this essay offers a few policy recommendations for individual states in order to arrest the downward spiral in East Asia's security predicament.

As to the nature and scope of this essay, a few caveats are in order. First and foremost is the geographical limitation to East Asia—that is Northeast and Southeast Asia, excluding China's northern and western neighbors. Another is the issue of Taiwan. It is certainly not a neighbor of China, especially in the eyes of Beijing; yet Taiwan is not only directly affected by China's military rise but it in and of itself is the driver for PLA modernization. Still another is that the naming and sequence of places or areas under dispute obviously do not indicate any preferences or hidden motives of this essay. They shall be avoided as much as possible.

Ⅰ. China's "Military Rise" in U.S.-led East Asian Security

Probably the most consequential aspect of China's "military rise" will be a change in the dynamics of power in East Asia, in which the U.S. maintains the leading and stabilizing role, a network of bilateral alliance and defense ties, and a set of economic and security objectives. It is thus no wonder that the possibility of power transition from the dominant U.S. to the rising China has attracted so much attention from the academic and policy community as well as from the international media.[2] Due also to the logic of great-power politics, the nature of the Chinese political system, and its continued involvement in the region's outstanding territorial and maritime disputes, it stands to reason that its neighboring countries are concerned about how China might use its new power and influence—now and in the future.

The realist-neoliberalist divide over the effects of China's ascendancy on U.S.-China relations and the possible power transition is rather well-known and will not be repeated here. Moreover, "[h]istorical parallels are by nature inexact," notes Henry Kissinger in his book *On China*. But he quickly asks whether or not World War I was caused by Germany's rise or by German conduct (and intentions). The answer is, as he alluded to a British foreign affairs official called Crowe in 1907, "it made no difference." What matters is an objective threat. "Whatever China's intentions," he continues, "the Crowe school of thought would treat a successful Chinese 'rise' as incompatible with America's position in the Pacific and by extension the world."[3]

The present-day "strategic distrust" between the U.S. and China is a case in point. The Chinese leadership perceives that the thrust of U.S. strategy toward the Asia-Pacific region and China is to contain China, so it needs to hold off America's encroachment as much as possible, whereas the Obama administration relies heavily—and increasingly so at the time of its federal

---

[2] It is important to note, however, that the power transition theory and its variants should be verified in terms of their theoretical logic and their contextual applicability. For two representative works on the critique of the theory, see Jack S. Levy, "Power Transition Theory and the Rise of China," in Robert S. Ross and Zhu Feng, eds., *China's Ascent: Power, Security, and the Future of International Politics* (Ithaca: Cornell University Press, 2008), pp. 11-33; Steve Chan, *China, the U.S., and the Power-Transition Theory: A Critique* (London: Routledge, 2008).

[3] The quotations and the Crowe episode are from the "Epilogue: Does History Repeat Itself? The Crowe Memorandum," Henry Kissinger, *On China* (New York: The Penguin Press, 2011), pp. 514-30, esp. pp. 518-20.

and defense austerity—on the linkage of alliance networks and friendly ties in the region.[4] The net effect is none other than "strategic access vs. strategic anti-access competition" at the regional level. Moreover, it needs only one side to make this vicious cycle tick. To wit, it does not always take two to tango. On this point Kissinger strikes a similar note.

> **China would try to push American power as far away from its borders as it could, circumscribe the scope of American naval power, and reduce America's weight in international diplomacy. The United States would try to organize China's many neighbors into a counterweight to China's dominance. Both sides would emphasize their ideological differences. The interaction would be even more complicated because the notions of deterrence and preemption are not symmetrical between the two sides. The United States is more focused on overwhelming military power, China on decisive psychological impact. Sooner or later, one side or the other would miscalculate.[5]**

A realist take or a "realistic" understanding of the rise of China is available from many of America's best minds. For one example, John J. Mearsheimer, assuming similar to Kissinger that unlike military capabilities state intentions are mutable and hard to divine, argues that "China will have aggressive intentions and will try to become a hegemon in Asia," thus causing a trouble for the U.S. who will prevent another "regional hegemon" (like itself) from arising in the world.[6] Another is Aaron L. Friedberg, who has

---

[4] For the impact of U.S. fiscal austerity on its rebalancing force readiness and defense institutions, see U.S. Office of the Secretary of Defense, *Quadrennial Defense Review 2014*, 2014.

[5] U.S. Office of the Secretary of Defense, *Quadrennial Defense Review 2014*, p. 521.

[6] For an excellent exposition on the history of great-power politics and its theoretical insight into the effects of a rising China on the region, see John J. Mearsheimer, "The Rise of China and the Fate of South Korea," paper presented at an international conference of "Korean Question: Balancing Theory and Practice" (Shilla Hotel, Seoul: Institute of Foreign Affairs

sized up China's rise for both cooperative and conflictive aspects but leaned toward differences with the U.S. in terms of national interests and ideological and political dimensions. While summing up the 1998 message by CIA Director George Tenet, Friedberg relayed that "China was a fast-rising power, determined to increase its influence, and likely some day to challenge America's preponderance in Asia and perhaps beyond."[7] For still another and at the official level, the annual DoD report to Congress on China's military developments, while echoing President Obama's January 2011 statement that "the United States welcomes a 'strong, prosperous, and successful China that plays a greater role in world affairs'," nonetheless notes that "Beijing is seeking to balance a more confident assertion of its growing interests in the international community with a desire to avoid generating opposition and countervailing responses from regional and major powers."[8]

Most, if not all, of the authors cited above would concur that at present China—with the exception of Taiwan and other sovereignty issues—is basically a status quo power in a sense that even if it is a rising power China is benefitting from the regional stability buttressed by the U.S., the world's sole superpower. In other words, China is basically "satisfied" with the U.S.-led regional security. The question is: for how long? For its part, the Chinese government often stated that the first two decades of the 21st century (i.e., 2000-2020) constituted the "period of important strategic opportunity" (*zhongyao zhanlue jiyu qi*) for its national development. Will China remain a

---

and National Security (IFANS), October 7, 2011); See also his speech text "Taiwan in the Shadow of a Rising China" (Taipei, Taiwan, the ROC: Annual Conference of the Association of International Relations, December 7, 2013).

[7] Aaron L. Friedberg, *A Contest for Supremacy: China, America, and the Struggle for Mastery in Asia* (New York: W. W. Norton & Company, 2011). The quotation is from p. 98. See also a critique of this book with an emphasis on the importance of the economic factor in their rivalry, Martin Jacques, "The Case for Countering China's Rise," *New York Times Book Review*, September 23, 2011.

[8] See Office of the Secretary of Defense, *Military and Security Developments Involving the People's Republic of China 2011*, 2011, pp. 1, 55.

status quo power even after it continues to rise in the 2020s and beyond? As illustrated in the above debate over "intention versus capability," a future China could behave like a revisionist state if it can—regardless of its intentions. This calls for an analysis of the means and ways China can employ to achieve its future status. One of the best cases that can be made is the PLA's anti-access strategy.

It is now well established that the term "anti-access/area denial strategy or A2/AD" originates in the PLA's preparations for a Taiwan contingency and that it aims at the approaching U.S. naval and air power in a crisis.[9] China's attempts to "deter, delay, and if possible defeat" the U.S. military intervention in a Taiwan contingency are supposedly based upon a combination of assets including a substantial submarine force, a fleet of fourth-generation aircraft, a variety of air-to-surface, ship-to-ship, and ballistic missiles with terminal guidance capability—i.e., anti-ship ballistic missile (ASBM),[10] and an array of coastal defense measures.

In PLA languages, on the other hand, there exists no terminology that corresponds to anti-access. Instead, such PLA strategies as "active defense"

---

[9] "Anti-access strategy" is a term coined by the U.S. military. First appeared in the 2001 *Quadrennial Defense Review* (QDR), it was mentioned more than 17 times in the version. See U.S. Office of the Secretary of Defense, *Quadrennial Defense Review*, 2010. For an authoritative exposition on the PLA's anti-access strategy, see Michael McDevitt, "The PLA Navy Anti-access Role in a Taiwan Contingency," paper presented at the NDU symposium of PLA Navy, <www.ndu.edu/inss>. See also U.S. Office of Naval Intelligence (ONI), *The People's Liberation Army Navy: A Modern Navy with Chinese Characteristics* (Washington, DC: Office of Naval Intelligence, August 2009).

[10] This capability is believed to be in the developmental stage and is aimed at striking aircraft carriers. Not surprisingly, it has been emphasized in the 2010 edition of *Military and Security Developments Involving the People's Republic of China*, especially p. 2, 29-30. See also Andrew S. Erickson and David D. Yang, "Using the Land to Control the Sea? Chinese Analysts Consider the Antiship Ballistic Missile," and Eric Hagt and Matthew Durnin, "China's Antiship Ballistic Missile: Developments and Missing Links," *Naval War College Review* 62, No. 4 (2009), pp. 53-86 and 87-117, respectively.

(*jiji fangyu*) or "strategic defense" (*zhanlue fangyu*) come close to the concept anti-access in effect. For example, in the authoritative *The Science of Military Strategy*, the authors underline that "The essence of the active defense is the offensive defense. Although strategic defense is defensive on the whole and passive in form, it is not purely defensive operations, and doesn't mean waiting passively for the enemy's attack."[11] It thus follows that at the operational level called "strategic maneuver" its major features consist of "Seizing the Initiative" (*zhangwozhudong*), "Acting Quickly" (*xingdongxunsu*), "Being Flexible" (*jidonglinghuo*), "Strengthening Coordination" (*jiaqiangxietiao*), "Combining Openness with Cover-up" (*gongkaiyuyinbijiehe*), "Combining Preposition[ing] with Maneuver" (*yuzhi he jidongjiehe*), and "Combining the Strategic Maneuver with Combat Operation" (*zhanluejidongyuzuozhanjiehe*).[12] In particular, while incorporating the changing features of modern local war, it even goes on to argue that "we should try our best to fight against the enemy as far away as possible [from China's borders/coast], to lead the war to enemy's operational base, even to his [*sic*] source of war, and to actively strike all the effective strength forming the enemy's war system."[13]

A gallery of the existing literature points to the particular set of military operations that intends to cope with a superior military power. They are: a) attacks on C4ISR systems including computer network attacks, EMP attacks, and attacks on satellites; b) attacks on logistics, transportation, and support functions; c) attacks on enemy air bases; d) blockades; e) attacks on sea lanes and ports; f) attacks on aircraft carriers; and g) preventing the use of

---

[11] The quotation is from the English version of the AMS's *ZhanlueXue*. See Peng Guangqian and Yao Youzhi, *The Science of Military Strategy* (Beijing: Military Science Publishing House, 2005), p. 307.

[12] These requirements are summed up in Peng Guangqian and Yao Youzhi, *The Science of Military Strategy*, pp. 318-20.

[13] Peng Guangqian and Yao Youzhi, *The Science of Military Strategy*, pp. 459-61. The quotation is on p. 461.

bases on allied territory. Complicated and daring they might be, they only appear to be the fuller spectrum of military actions and capabilities that have potential access-denial effects; the level of operational advancement in each category—let alone their actual performance in conflict—will be all but different from each other.

Most PLA analysts would agree that the PLA has gradually but considerably improved its fighting capability for nearly 30 years through its across-the-board defense modernization. Notwithstanding a slow yet steady progress for the first 15 years (1985-1999) in a wide array of areas such as organization, equipment, training, and defense industry, a glaring weakness remained as to military technological backwardness, outmoded and obsolescent weapons systems, inadequate number of high-tech systems, and a lack of realistic training. The post-Tiananmen western embargoes as well as Sino-Soviet/Russian rapprochement (in particular, the decline of the Russian state power) allowed PLA leaders to seek from Russia high-tech weapons and technologies. Parenthetically, the 1995-96 crisis in the Taiwan Strait and the subsequent U.S. military involvement further reinforced China's military-equipment dependence on Russia.[14]

In particular, since around 2000 a new and more extensive pattern in PLA force modernization has been observable from outside. They include, but are not limited to, a) the production and deployment of new weapons systems; b) introduction of more and better weapons systems from abroad, particularly from Russia; c) enhanced rapid reaction capability (RRF); d) a steady increase in information warfare (I/W)/ information operation (I/O), and EW capability; e) improved integrated logistics system (ILS); f) widespread "joint" MR training and exercises; and g) production and deployment of a variety of missile systems.

---

[14] The most comprehensive account in Chinese of the 1995-96 Taiwan Strait crisis is Qi Leyi, *Hanweixingdong: 1996 Taihaifeidanweijifengyunlu* [Defense Action: Records of the 1996 Taiwan Strait Missile Crisis] (Taipei: Liming wenhua, 2006).

China's import of Russian weapons and technologies, for instance, jumped to more than two billion dollars from the previous one billion dollars per year. In addition to new hardware (such as the naval Su-30MK2, Il-76 [for transport and AWACS], and Il-78 [for mid-air refueling]), their technological cooperation includes parts, design, R&D, and operational know-how. It is in this context that China became the world's largest importer of major conventional weapons in the period from 2001 to 2008.[15] China's "indigenous" development of HQ-9 and FT-2000 SAMs may also have been aided by the import of a whopping 994 S-300PMU/SA-10 missiles and its related technologies by 2006.[16]

It is thus not surprising to note that China's acquisition of Russian weapons and technologies has been *the* most important source for PLA's new "indigenous" weapons development and force modernization as well as by extension for its anti-access capability.[17] The current and likely future acquisition processes invariably point to a continued acquisition of Sukhoi (Su) family aircraft, J-11A/B (China's licensed or unauthorized product of Su-27SK), and J-10 (China's domestically developed combat aircraft); KJ-2000/KJ-200 AWACS; Ilyushin (Il) series transport (Il-76) and tanker (Il-78) aircraft and its domestic variants (Y-20); and engines (WS-10), radar, design, and avionics. There also is a whole range of new air-to-air and air-to-surface missiles with an eye on anti-ship mission.[18] In brief, it is

---

[15] See *SIPRI Yearbook 2009*, p. 326.

[16] U.S. Office of the Secretary of Defense, *Military Power of the People's Republic of China 2006*, 2006, p. 21 and the author's estimates.

[17] For a succinct and balanced assessment of Russia's role in China's force modernization, see *SIPRI Yearbook 2009*, pp. 308-10. This author, on the other hand, has long argued for a distinction between a fighting capability and force modernization. They are simply different concepts with each other.

[18] For a comprehensive assessment of the current status of PLAAF modernization, see Craig Caffrey, "China's Military Aircraft: Up and Coming," *Jane's Defence Weekly*, July 5, 2010;

highly likely that the thrust of future PLA force modernization—at least until 2020—would very much follow the course identified in the first decade of the 21st century.[19]

It is thus necessary to keep an eye on the Russian assistance and transfer to China of weapons systems and military technologies. As illustrated earlier, Russia has been the most important source for the PLA's advanced weapons and platforms—and by extension for its anti-access capability. For instance, while China's sea denial capability has similarly been buttressed by the introduction of new types and classes of missile destroyers and frigates as well as new submarines for the past two decades or longer,[20] it is *Sovremenny*-class destroyers and *Kilo*-class submarines that are most capable, thus being assigned to the highest mission-critical areas. Finally, it is noteworthy that Sino-Russian military cooperation began to resume in October 2010 after a three-year hiatus—in the form of visits to Russia by ranking Chinese officials.[21]

Kenny Fuchter, "Air Power and China in the 21st Century," *Air Power Review*, Vol. 11, No. 3 (Winter 2008), pp. 1-17.

[19] It should be noted, however, that the so-called "non-equipment" aspects of PLA modernization such as leadership, C4ISR, NCW, and joint exercises have not been addressed in this paper.

[20] For a comprehensive analysis of the PLAN's history, mission, and equipment, see Bernard D. Cole, *The Great Wall at Sea: China's Navy in the Twenty-First Century*, 2nd ed. (Annapolis, MD: Naval Institute Press, 2010). See also Ronald O'Rourke, *China Naval Modernization: Implications for U.S. Navy Capabilities*, Background and Issues for Congress (Washington, DC: CRS, November 2009). This report has been updated several times since its first publication. The PLA Navy has operated the Romeo, Golf, Whiskey, Ming, Song, and Yuan subs. Destroyers include Luda, Luhu (Type 052), Luhai (Type 051B), Luyang I/II (Type 052B/C), and Luzhou (Type 051C).

[21] They include CMC Vice Chair Guo Boxiong (September 2011), Chief of General Staff Chen Bingde (August 2011), State President and CMC Chair Hu Jintao (June 2011), State Council Vice Premier Li Keqiang (April 2012), and State President and CMC Chair Xi Jinping (March 2013).For a brief overview of recent visits, see Jingdong Yuan, "Sino-Russian Relations: Renewal or Decay of a Strategic Partnership?" *China Brief*, Vol. 11,

All in all, China's acquisition of advanced weapons and military technologies from abroad since 2000 is geared to establishing a modern fighting force. Its acquisition patterns also indicate the PLA's continuing difficulties as well as its future direction of "army building." Even if most of the imported and license-produced assets can be assigned to various anti-access role, they are not a good measure of the PLA's actual war-fighting capability, however. Not only does it take years of practice and training with new equipment but the theater of possible conflict often makes a big difference in performance, as shown below.

II . The Regional Context: "Getting the Questions Right"

In assessing the regional implications of PLA's force modernization and China's military rise, it is very important to understand that it is not China's defense modernization per se, but its actual and perceived capability to project and sustain power along and beyond its borders that has a direct bearing on achieving its foreign policy goals and arouses concerns over its capability to destabilize regional security. The angst will continue as long as there remains a possibility that China would try to resolve forcefully a host of outstanding disputes with its neighboring countries.

Of great relevance to this study is, for one thing, East Asia's political and geographical diversity. Land-based inter-Korean confrontation, for example, requires an approach quite different from disputes in the South China Sea, which are maritime and multilateral in nature. Besides, in Northeast Asia when compared with Southeast Asia there exist a higher level of hostility and militarization, bilateral nature of conflict, and the lingering impact of bilateral security arrangements. This partly explains the difficulties of establishing viable multilateral security arrangements, notwithstanding the SCO, Six-Party Talks, and the NEA Summit Meeting.

Issue 18 (September 30, 2011), pp. 11-14; Stephen Blank, *Shrinking Ground: Russia's Decline in Global Arms Sales* (Washington, DC: The Jamestown Foundation, 2010).

Another is, as noted at the outset, China's neighbors' overall lower level of military capability than China's. There is little value in asking when and how the PLA will catch up the U.S. military. Nor is it useful to engage in "bean counting"—that is, enumerating each side's order of battle or force-on-force comparison. While PLA Navy, for instance, is weak in air defense, sea-based air power, and ASW capability, many of its neighbors' navies are weaker still. It calls, among others, for an individual country's own strategic and military assessment on the implications of China's military rise.

Still another is that for now and for the foreseeable future, the PLA Navy would continue to pursue the twin goals of preparing for a Taiwan contingency as well as of building a truly regional navy.[22] Given China's expanding national interests—and the growing role of the navy and maritime issues in them, it is commonsensical. They now range from out-of-area anti-piracy operations (such as those in the Gulf of Aden) to international disaster relief activities (HA/DR) and from SLOC and seaborne trade/energy protection to UN peacekeeping role.[23] While many of the military requirements for both goals do overlap, it is prudent to treat them as an integral whole.

---

[22] While a Taiwan contingency may be the main driver of current PLA modernization, its scope and nature strongly indicate that it also aims at achieving goals beyond Taiwan. See Roy Kamphausen, David Lai, and Andrew Scobell, eds., *Beyond the Strait: PLA Missions Other than Taiwan* (Carlisle Barracks, PA: U.S. Army War College Strategic Studies Institute, April 2009); Michael A. Glosny, "Getting Beyond Taiwan? Chinese Foreign Policy and PLA Modernization," *Strategic Forum*, SF No. 261 (January 2011),<www.ndu.edu/inss>.

[23] For such overseas activities in 2009-2010, see *China's National Defense 2010*.

III. Southeast Asia/The South China Sea: Geography and Limited Power Projection

China does need and has worked for a peaceful and stable external environment—the foremost reason being for its own economic development. As to the disputes in the South China Sea, it has long called for dialogue and diplomatic negotiation—mostly on a bilateral basis, signed the 2002 Declaration on the Conduct of Parties in the South China Sea, and actively participated in various ASEAN forums such as ARF, ASEAN+1, and ASEAN+3. A former PLAN admiral even asserts that "[F]ollowing the extensive growth of China's influence and power in Asia, the apprehension and doubts of neighboring countries toward China have not increased but rather decreased."[24]

Even if they were to be taken at face value, it is also true that China has continued to build up its military capability, especially its naval and air assets. A flurry of recent events, such as fishery and maritime disputes with Vietnam and the Philippines, the expansion of naval ports and airfields on Hainan Island, and an increase in maritime patrol activities, are not reassuring to its ASEAN neighbors. While many have argued that China's foreign-policy behavior took a more "assertive" turn in the 2009-2013 period, it is wiser to keep in mind that China has almost always been assertive and aggressive when it comes to what China defines as "sovereignty and territorial issues"—now coupled with the "core interests" (*hexinliyi*).[25]

---

[24] The quotation is from Yang Yi, former director of NDU's Institute for Strategic Studies, "Navigating Stormy Waters: The Sino-American Security Dilemma at Sea," *China Security*, Vol. 6, No. 3 (2010), pp. 3-11, esp. p. 5.

[25] On the other hand, China did resolve territorial disputes through diplomatic settlements in the early 1960s, when its counterparts accepted two principles—that is, non-recognition of "unequal treaties" and the maintenance of the status quo. They include Nepal (1961), North Korea (1962), Mongolia (1962), Pakistan (1963), and Afghanistan (1963). In the 1990s as well China employed non-military solutions to a host of territorial disputes including Laos (1991), Russia's western border (1994), Kazakhstan (1994), Kyrgyzstan (1996), Tajikistan

Given the existing gap in military capability between China and its ASEAN neighbors, it would be a herculean task for the latter to cope with a small, isolated contingency should it arise. Their overall weaknesses in military preparedness, especially those of Vietnam and the Philippines, are well known.[26] On the diplomatic front, furthermore, the ten ASEAN member-states are deeply divided over the Spratly issue: Laos, Cambodia, and Myanmar (formerly Burma) remain sympathetic to China; Malaysia and Indonesia are wary of U.S. involvement in the row; Thailand and Singapore mostly take a neutral stance; and Vietnam and the Philippines, as claimants, are most vocal about China's stance and actions.[27] This again points to the continued importance of U.S. presence in maritime Southeast Asia. In brief, China's stated emphasis on regional stability and a "responsible great power" (*fuzeren de daguo*) is one thing, and its realpolitik behavior based on hard-nosed national interests could be quite another.

(1999), Vietnam (1999), and Russia's eastern border (2004), mainly for a stable and peaceful external environment conducive to domestic economic development. For details, see Taeho Kim, "China's Territorial Ambitions? Enduring Patterns and New Developments," in the Korea Institute for Maritime Strategy (KIMS), ed., *PLA Navy Build-up and ROK Navy-US Navy Cooperation* (Seoul: KIMS, 2009), pp. 97-113 and 339-57 in Korean and English. On China's core interests and its assertive behavior Michael D. Swaine has over the years offered a detailed and perspicacious analysis on their various aspects. See his "Part Four: The Role of the Military in Foreign Crises," *China Leadership Monitor* (hereafter abbreviated as *CLM*), No. 37 (April 30, 2012), <www.hoover.org/publications/china-leadership-monitor>; "Part Three: The Role of the Military in Foreign Policy," *CLM*, No. 36 (January 6, 2012); "Part Two: The Maritime Periphery" (with M. Taylor Fravel), *CLM*, No. 35 (September 21, 2011); "Part One: On 'Core Interests'," *CLM*, No. 34 (February 22, 2011) and "Perceptions of an Assertive China," *CLM*, No. 32 (May 11, 2010).

[26] See, for example, Evan S. Medeiros, "The New Security Drama in East Asia: The Responses of U.S. Allies and Security Partners to China's Rise," *Naval War College Review*, Vol. 62, No. 4 (2009), pp. 37-52.

[27] See, for example, Trefor Moss, "Regional Matters: Regional Overview—Southeast Asia," *Jane's Defence Weekly*, May 6, 2011. For a recent analysis, see Darshana M. Baruah, "South China Sea: Beijing's 'Salami Slicing' Strategy," *RSIS Commentaries*, No. 054/2014 (March 21, 2014).

It is thus the function of U.S. presence in the theater—in the form of overseas bases and the freedom of navigation—and the PLA's own limitations to project and sustain power for an extended period of time that have largely prevented armed conflict from occurring for the past decade or so. Geography alone makes the PLA's power projection a difficult proposition, including long-range air and naval assets (e.g., refueled Su-30MK2s, *Sovremenny*-class destroyers, and *Kilo*-class subs), replenishment-at-sea (RAS) capability, ocean surveillance satellites, and even carrier-based air power. Many of these at-sea requirements are being acquired by the PLA. Until the full constellation of forces is in place, however, the PLA's anti-access operations against the U.S. military will be limited to the areas that are within the protection range of land-based assets.[28]

While the Southeast Asian states' military preparations are no match to those of China—in terms of their quantity and quality, China as well as its involvement in the South China Sea dispute is a factor in this sub-region's military modernization.[29] One of the most notable developments includes the acquisition of submarines: six Kilo-class Type 636 submarines by Vietnam; three Type 209/1400 submarines by Indonesia; and possibly four

---

[28] For a succinct review of PLA's limitations in projecting power in the theater of the South China Sea, see Bernard F.W. Loo, "Chinese Military Power: Much Less than Meets the Eye," *RSIS Commentaries*, No. 111 (September 9, 2010), pp. 1-2.

[29] For a comprehensive yet detailed discussion of selected Southeast Asian countries' defense build-up, see Trefor Moss, "Regional Matters: Regional Overview—Southeast Asia," *Jane's Defence Weekly*, May 6, 2011. In March 2014 Indonesia announced that it had overlapping claims to several islands with China, thus identifying itself as a claimant to the dispute. See Ann Marie Murphy, "The End of Strategic Ambiguity: Indonesia Formally Announces Its Dispute with China in the South China Sea," *CSIS*, PacNet #26 (April 1, 2014). In the same month the Philippine government and the Moro Islamic Liberation Front (MILF) signed a peace agreement, thus ending the 44-year-long internal armed struggle.

Vastergotlands submarines by Singapore.[30] New combat aircraft acquisitions are notable as well: 12 FA-50 light attackersby the Philippines,[31] and probably 12 Su-30s by Vietnam.

IV. The Taiwan Strait/Taiwan: Still the Leitmotif but a Tough Nut to Crack

The Taiwan issue is the *idée fixe* of China's diplomacy and military. In order to conduct a successful military campaign in a Taiwan Strait contingency, the PLA will need to have air superiority, sea control, amphibious capability, and missile strikes. It is sober to note that the PLA has achieved significant progress in each category over the years. In particular, the PLAAF and PLANAF's acquisition of modern combat aircraft is quite impressive, especially seen against Taiwan's eroding air defense capability.[32] Its sea control capability has similarly been buttressed by the introduction of new types and classes of missile destroyers and frigates as well as of 38 new submarines in the 1994-2007 period.[33] Its amphibious capabilities, though less noticed, have grown in numbers and loading capacity as well.[34] Finally,

---

[30] For new developments on sub acquisitions, see Koh Swee Lean Collin, "Vietnam's New Kilo-class Submarines: Game-changer in Regional Naval Balance?" *RSIS Commentaries*, No. 162/2012 (August 28, 2012) and his "Indonesia's New Submarines: Impact on Regional Naval Balance," *RSIS Commentaries*, No. 021/2012 (January 2012).

[31] See *Dong-A Ilbo*, March 28, 2014; *Chosun Daily*, March 28, 2014.

[32] Probably the most cogent often-source argument to shore up Taiwan's air defense capability with the acquisition of F-16C/D Block 50/52 is Lotta Danielsson-Murphy, ed., *The Balance of Air Power in the Taiwan Strait* (Arlington, VA: US-Taiwan Business Council, May 2010).

[33] For a comprehensive analysis of the PLAN's history, mission, and equipment, see Bernard D. Cole, *The Great Wall at Sea: China's Navy in the Twenty-First Century*, 2nd ed. (Annapolis, MD: Naval Institute Press, 2010).

[34] See James C. Bussert and Bruce A. Elleman, *People's Liberation Army Navy: Combat Systems Technology, 1949-2010* (Annapolis, MD: Naval Institute Press, 2011), p. 98, Table 24 and Richard D. Fisher, Jr., *China's Military Modernization: Building for Regional and Global Reach* (Stanford: Stanford University Press, 2010), p. 154, Table 6.8.

the PLA's growing inventory of missiles—in the form of SSM, LACM, and ASCM—pose a distinct threat of and by itself and especially when combined with other assets.[35]

The operational and tactical measures the PLA can employ are extensive and highly complicated. But some of the most plausible—thus most effective—approaches are identifiable. For one thing, air superiority will be indispensable for the follow-up naval and amphibious operations and for weakening Taiwan's defense capability. In a scenario intended to conclude the conflict before the third party's intervention, the airpower can be particularly instrumental. For another, a wide array of ballistic missiles in the PLA's possession could inflict damage to many valued targets in Taiwan such as command and control centers, airfields, naval ports, and other military installations. Missile attacks could precede or could be conducted in tandem with air strikes to do great damage; but they alone cannot disable Taiwan's ability to counterattack. Nor is it sufficient to foil the will to fight of Taiwan people. For still another, some of the so-called "non-equipment" aspects such as leadership, C4ISR, NCW, inter-service coordination are essential ingredients for a successful campaign over the Taiwan Strait.

Timely arrival of three to four U.S. carrier battle groups (CVBs)—which is a big if—in and near the Taiwan Strait, together with sufficient expeditionary forces, would lead to a failure in PLA's anti-access efforts and may even thwart the invasion itself. Though extremely complicated, a numerical count indicates that in the case of PLAAF's anti-ship role a total of one JH/FB-7 regiment (20 to 22 aircraft), two Su-30MKK regiments (44 to 48), one Su-30MK2 regiment (22 to 24) and older bombers (H-6D) can be

---

[35] See, among others, Eric C. Anderson and Jeffrey G. Engstrom, *Capabilities of the Chinese People's Liberation Army to Carry Out Military Action in the Event of a Regional Military Conflict* (McLean, VA: Science Applications International Corporation, March 2009), pp. 48-49, 52.

assigned.[36] They are not, in brief, sufficient in numbers to face the carrier-based airpower and air defense. For this reason, the PLA has worked on developing longer-range ALCM and ASBM.

Similarly, the PLAN's substantial submarine force remains a main pillar of the sea denial mission. If deployed to the boundaries of sea-denial area, a minimum of six subs are needed per each carrier battle group. It thus needs 18 to 24 subs for three to four CVBs; yet for a rotation of at-sea operation, en-route to port, and maintenance and resupply at port it will require a total of 54 to 72 submarines. At present, the PLAN does not possess such numbers, even if it is likely to grow in numbers down the road. Besides, lack of practical experiences, logistics training, and operational know-how will be difficult hurdles to surmount for a successful PLAN sub operation.

Beyond the specific focus on the military dimension, much will depend on the future of the current cross-Strait rapprochement. Will the combination of "a rising China, a weakened Taiwan, and declining U.S. support," as Robert Sutter has argued, push Taiwan further into China's orbit?[37] Or would the shift in Taiwan's domestic debate from identity to economy ensure the continuation of the Taiwan-China détente and integration?[38] The two views see the different futures of the cross-Strait relations: the former implies an inevitable change in the status quo, whereas the latter presupposes the status quo as the only realistic alternative. At least in the former case,

---

[36] They are the author's own estimates. See the recent and different counts made by Kenneth Allen, Mike McDevitt, and Anderson and Engstrom. For PLAAF's recent order of battle, see *The Military Balance 2014*, pp. 235-236.

[37] Robert Sutter, *Taiwan's Future: Narrowing Straits* (Washington, DC: NBR Analysis, May 2011). See also Parris H. Chang, "Beijing Copes with a Weakened Ma Administration: Increased Demands, and a Search for Alternatives," *China Brief*, Vol. 14, Issue 2 (January 17, 2014).

[38] The argument has been made by Lowell Dittmer, "Taiwan's Security in an Era of Cross-Strait Détente," paper presented at the "18th Smart Talk Seminar" (Seoul: East Asia Institute (EAI), July 15, 2011).

which envisions the deepening of the current trajectory of Taiwan's integra-tion with China, the U.S. will become less relevant—with unknown conse-quences to regional stability.

V. Northeast Asia/The East China Sea, The Korean Pen-insula: Intermix of Geostrategy and History

The future of East Asian prosperity and security will be largely shaped by the economic and security trajectories of China and Japan as well as by U.S. relations with both countries. A continued healthy U.S.-Japan security relationship is vital to American interests and to Asian security, and for the moment the U.S. has a felt need to support a "normal Japan," but without jeopardizing its neutral stance on historical and territorial issues. In particular, a series of recent statements by top Japanese leaders that provoke its neigh-bors is causing annoyance in U.S. East Asia strategy and its alliance man-agement. At the same time, it is very important to maintain the position that the alliance should not appear aiming at China.

China and Japan embody the world's second and third largest econo-mies, respectively, and wield substantial political clout in regional affairs. Militarily, albeit different in nature and size, both countries are major factors to be reckoned with in any East Asian strategic equation. Thus, the current spate of antagonistic ties between Beijing and Tokyo—as well as those be-tween Tokyo and Seoul, an ally of the U.S.—remains problematic. On these two sets of bilateral ties, the U.S. has so far been unable to resurrect normal ties, while continuing to emphasize that the most pressing agenda should be security, not history.

As noted above, while the future of China-Japan relations will have a substantial impact on post-Cold War East Asia's economic and security order, their traditional rivalry and current and likely future power potentials will continue to be a source for concern in their neighbors' strategic planning. For both historical and contemporary reasons, each country has also pursued its foreign policy goals with an eye on the other.

In terms of future regional stability, what is perhaps more significant in the beginning of the new century is whether the two major regional powers will develop a relationship that is either strong and cooperative *or* weak and confrontational in the years ahead. Of equal importance is the diverse yet uncertain impact of this evolving relationship on the future of East Asian security, particularly in light of their changing domestic and international contexts.

As China's continued economic growth depends more on securing maritime resources and interests, it stands to reason that the PLA Navy will acquire a wider range of mission capabilities. This type of naval modernization is bound to enhance the level of apprehension by other regional powers and even create an action-reaction cycle at sea. While the current discussion on this subject tends to focus on the U.S.-China rivalry, an important yet under-researched aspect is the creeping regional power transition between China and Japan—traditional regional heavyweights. A combination of "resistant nationalism," a sense of crisis, political immobility—especially by the post-war generation of political leaders—is sweeping over Japanese society.[39] While the ongoing power transition between the two major regional powers will remain an issue of greater attention.

As perhaps best illustrated by the abrupt yet enduring controversy between Beijing and Tokyo over the Senkaku/Diaoyu/Diaoyutai Islands since the late summer of 2012,[40] neither side wants to appear meek on such sensi-

---

[39] "Resistant nationalism" is the author's understanding of Waseda University Professor Lee Jong Won's discussion on Japanese tendency on nationalism and statism. See an interview with Professor Lee, *Dong-A Ilbo*, September 24, 2012.

[40] For a detailed analysis of the island dispute, see Michael D. Swaine, "Chinese Views Regarding the Senkaku/Diaoyu Islands Dispute," *China Leadership Monitor*, No. 41 (June 6, 2013); Paul J. Smith, "The Senkaku/Diaoyu Island Controversy: A Crisis Postponed," *Na-*

tive and nationalistic issues, due in part to domestic imperatives. New leadership line-ups—President Xi Jinping and Prime Minister Shinzo Abe—have their own agenda to sustain the tense atmospherics in their bilateral ties. China's announcement of its Air Defense Information Zone (ADIZ) in November 2013 is only the latest manifestation of this deeper and difficult relationship between the two major regional powers.

It is argued finally that despite their huge and growing stakes in maintaining an amicable relationship, China-Japan relations will remain a difficult and often tense process. The persistence of their traditional rivalry and historical distrust over time suggests that they may have more to do with deeply ingrained cultural, historical, and perceptual factors than with the dictates of economic cooperation or shared interest in regional stability that are mutually beneficial. Also underlying their complex but competitive ties is the rise of new-generation leaders in both countries who are tasked with coping with a complex set of challenges from below as well as from outside. How well and in what manner they handle the challenges could significantly affect not only the wealth and health of their respective nation but also the future of the regional order. The future stability in East Asia will hang in the balance as China and Japan continue to seek a new balance between their interdependence and rivalry.

That South Korea (or the ROK) and China have since 1992 remarkably improved their bilateral relations in all major issue-areas is beyond doubt.[41]

---

val *War College Review*, Vol. 66, No. 2 (Spring 2013), pp. 27-44; Noboru Yamaguchi, "A Japanese Perspective on the Senkaku/Diaoyu Islands Crisis," in The East Asia Program, ed., *Tensions in the East China Sea* (Sydney: The Lowy Institute for International Policy, December 2013), pp. 7-17; and Barthelemy Courmont, "Territorial Disputes and Taiwan's Regional Diplomacy: The Case of the Senkaku/Diaoyu/Diaoyutai Islands," *Journal of Territorial and Maritime Studies* (Seoul), Vol. 1, No. 1 (Winter/Spring 2014), pp. 113-134.

[41] For a brief account of Sino-ROK relations as well as the gap between rhetoric and reality, see Taeho Kim, "Sino-ROK Relations at a Crossroads: From *Qiutongcunyi* (求同存異) to *Yizhongqiutong* (異中求同)," *New Asia*, Vol. 19, No. 2 (Summer 2012), pp. 34-44.

Yet, the ROK and China have put an uneven emphasis on economic and so-cio-cultural relations.[42] In the political and diplomatic fronts their percep-tions often diverge from each other—as vividly seen in the aftermath of the *Cheonan* sinking (March 2010) and the artillery shelling of the Yeonpyong Island (November 2010). For its part, China appears indifferent to such sen-sitive yet important issues as North Korean contingencies, the history of Koguryo, and the plight of North Korean residents in China.

China's "military rise" is an issue of growing security concern to the ROK, but it is often viewed as of long-term nature.[43] Of all factors that af-fect the ROK's calculations the geostrategic and historical considerations remain most enduring and consequential. First, the peninsula is not only lo-cated closest to China's capital but also shares a 1,400-kilometer (880-mile) land border with it. Furthermore, Chinese strategists often regard the penin-sula as a "route" (*tonglu*) between the maritime and continental powers. Second, it is also in this peninsula that the fledgling PRC fought with the U.S. 60 some years ago. Before that, historical rivalry between China and Japan over the peninsula and the West Sea (Yellow Sea) also illustrates the strategic importance of the peninsula. Third and in China's view, the fast growing economic ties between Beijing and Seoul testify the vicissitude of Cold-War politics and the validity of China's ongoing reform and opening drive. Fourth, not only was traditional Korea part of the Sinocentric world

---

[42] The other major aspects of 22-year ties between the ROK and the PRC include: a) uneven growth in different issue-areas; b) rapid expansion in the number of actors and in the scope of their ties; c) the effect of the "rise of China" on their bilateral ties; and d) the growing gap in their respective national power.

[43] How long the North Korean threat will last is an intriguing yet unanswerable question. At the time of this writing in July 2014 North Korea launched several different types of mis-siles and MLRSs toward the East Sea—together with the threat of fourth nuclear test, about 100 rounds of artillery shells within South Korea's Northern Limit Line (NLL) in the Yel-low Sea, and sent an unidentified number of drones into the South for reconnaissance pur-poses.

order but China's potential to become a full-fledged great power will likely be tested again in this peninsula.

More specifically, China's operational SSNs and SSBNs are not only harbored in the North Sea Fleet and mostly patrol in the West Sea (Yellow Sea) and East China Sea.[44] China's future carrier battle groups, once they become operational, would also likely be located in the vicinity of the peninsula. China's increasing number of modernized combat aircraft as well as of conventional missiles needs to be reckoned with, even if they are not necessarily targeted at the peninsula. More immediate attention should be given to the PLA's RRFs. By the present estimate, seven out of the PLA's 18 group armies (GAs) are RRUs or mobile forces (MFs), of which four are located in the Beijing (38th and 27th), Shenyang (39th), and Jinan (54th) MRs. In light of the past patterns of China's use of force in a diplomatic crisis as well as the growing body of evidence for North Korea's internal weaknesses, they could be employed in a variety of North Korean contingencies such as humanitarian cases, a large flow of refugees, and instabilities in the border areas.

South Korea is a genuine middle power by any definition. Given its geographical location as well as its neighboring major powers, on the other hand, it is a relatively weaker power. To overcome its continuing plight, there are only two ways: "internal balancing" or "external balancing." The object of the latter should have a) no territorial ambitions; b) a will and capability to assist in time of crisis; and c) a proven historical record to be a benign power. That the only country which meets the three conditions is the United States is a grim reality. Besides, it is imperative to ponder over that how South Korea emerged as a major economic power with its enhanced international stature during the Cold War. The essence of its external balancing is therefore to maintain a rock-solid relationship with the United States.

---

[44] The annual DoD report on China's military estimated that there are three SSNs in the North Sea Fleet. See U.S. Office of the Secretary of Defense, *Annual Report to Congress: Military and Security Developments Involving the People's Republic of China*, 2012, p. 31.

China's "military rise" will continue to influence the current ROK and future Korea's security environment. In addition to military consideration, therefore, the ROK should work for the improvement of overall bilateral ties and pave the way for an eventual unification. As long as China's future positions and role in the peninsula remain uncertain, the ROK must simultaneously pursue toward China both "exchange and cooperation" and "anticipation and preparation" in case China changes its current course of "peace and development." A hedging strategy will remain the most reasonable approach for the foreseeable future.

On the military side the ROK's force modernization based on the principle of "limited defense sufficiency" should continue.[45] It means, among others, a minimum defense capability to deter and deny military provocations and to respond to small-scale conflict on and near the peninsula. In the near term, it should be able to cope with maritime conflict on top of the existing military threats from North Korea. In the mid- and longer term it calls for a capability to deny or raise the cost of military provocations, which depends upon a more independent intelligence-gathering capability (e.g., E-737), effective naval and air power (Types 209/214, F-15K), and a high-tech force. When and if China's "benign and reliable" policy is not forthcoming and in particular it becomes a more dominant power with a campaign-level fighting capability, the ROK cannot but further strengthen its defense ties with the United States.

An intriguing question is whether closer ROK-U.S. alliance relations would invite a harsher reaction from China—thus detrimental to the ROK's relations with China—or they would be beneficial for China's overall posture toward it. As long as the Chinese government views the alliance as part

---

[45] This is the author's term, which apparently is congruent with the ROK government's new post-*Cheonan* defense posture called "proactive deterrence." The latter term seems more targeted at the North Korean threats and has yet to be incorporated in the full report, which is in the making at the time of this writing.

of U.S. "containment" strategy and its position remains similar to that of North Korea, China is likely to take a critical stance toward the alliance, making peninsular and regional issues more difficult to resolve. The opposite—i.e., the ROK's distraction from the alliance—would be far more consequential for the ROK, however, possibly leading to a more independent yet isolated state without a reliable ally. It is, thus, in the interest of the ROK to maintain rock-solid ties with the United States, notwithstanding the rise of China or their effect on the U.S.-China relationship.

VI. Some Concluding Observations

From the above analysis on China's "military rise" and their region-wide implications, it can be deduced that the effectiveness of the PLA's anti-access strategy, originally designed for a Taiwan Strait crisis, is equally applicable to other regional contingencies. The Spratly disputes and a North Korean contingency are cases in point. In so far as China's anti-access strategy intends to deter and delay the arrival of U.S. naval and air assets and the U.S. in such contingencies must use the forward bases in the region, most regional actors will not be immune from the effects of the PLA's anti-access capabilities. For some of China's neighbors this means the need for developing the "mini anti-access strategy" of their own.

All in all, it is imperative that the PLA's force modernization and its anti-access efforts be assessed on a regular and objective basis. China's future force build-up will also be a function of mixed factors such as Chinese leaders' perceptions of its own security environment, the availability of domestic and foreign sources, and internal/bureaucratic constraints. Additionally, future trends in China's defense resources allocation—which will be largely affected by the current 12th Five-Year Plan (2011-15)—would be a good indicator for its trajectory of enhanced anti-access capability and/or other priority missions. Without greater transparency on the part of the PLA, outside analysts cannot help but work with the trends and outcomes.

The regional actors may well be advised to think hard about the military implications of China's ascendancy to themselves and to the regional balance of power. Multilateral security fora are just one such endeavor, which by all means should be encouraged; yet many salient sovereignty and territorial issues are addressed at the bilateral, not multilateral, level. The multi-layered competition between the U.S. and China at the strategic, military, and access/anti-access level also calls for individual regional actors to cooperate and coordinate with each other in peacetime. In brief, a hedging strategy buttressed with a web of multilateral networking will prove to be most prudent for years to come.

Finally, regional actors should be able to read correctly the changes in East Asia's security environment. Not only are they mostly nations with democratic governance and market economies, they also share a set of common goals—in real and idealistic terms. In addition, they are in need of preparing for the present and likely future challenges to their own security and they also are intertwined with each other on a panoply of such security issues as missile defense (MD) systems, anti-access/area-denial capability, and maritime safety—to name but a few. They will be the most cost-effective investment for continued peace and prosperity for all individual countries and the region as a whole.

Japan's Security Strategy toward the Rise of China From a Friendship Paradigm to a Mix of Engagement and Hedging

Tsuneo "Nabe" Watanabe
(Senior Fellow, Tokyo Foundation)

History is an essential component of Japan's strategy toward China. Japan owes both its positive and negative legacy to its longstanding ties with China. Throughout history, Japan has been influenced by Chinese civilization, importing a system of writing, a structure of government, and ways to prepare food. Separated by the sea, Japan was able to maintain political independence from powerful Chinese dynasties through the centuries. Japan had created a modern nation-state by the late nineteenth century, while China became subject to colonization by European imperial powers. The United States has played a unique role in both Japanese and Chinese history. The country pressured Japan to open up its closed ports despite great reluctance after over 200 years of a *sakoku* seclusion policy. A rapidly modernized Japan sought to join the ranks of imperial powers by attempting to colonize parts of China.

The United States eventually crushed Japan's imperial ambitions in East Asia and secured China's independence. After World War II, the United States chose Japan as its closest ally in East Asia under a Cold War structure. China occasionally voiced its acceptance of the Japan-US alliance in strategically calculated moves to counter the Soviet Union or to secure its own economic development. The alliance with the United States is today the cornerstone of Japan's security strategy. At the same time, frustration has occasionally been expressed toward the United States, often accompanied by calls for a "return" to closer rapport with China.

Japanese leaders often note the importance of keeping the Japan-US-China relationship an equidistant one.[1] Such an attitude may be a reflection of the ambivalence in the Japanese psyche, at once contrite over past aggressions toward China and frustrated with the current overreliance on the United States.

Eric Heginbotham and Richard Samuels have described Japan's current policy direction as a dual hedge strategy, pursuing security interests through an alliance with the United States and economic interests through trade with China.[2] It has also been one containing many contradictions and which has sent unintended negative messages to both Washington and Beijing.

This paper will explore Japan's strategy toward China within the context of the highly complex Japan-US-China trilateral relationship, focusing on the historical legacy and balance of power, as well as the influence of domestic political dynamics. Japan's grand strategy can be said to have been shaped by the pursuit of complex security and economic interests within the Japan-US-China triangle, whose history can be traced back to the 1930s. The paper, though, will limit its scope to the period after 1972, when Japan normalized diplomatic relations with the People's Republic of China, to the present, when Japanese strategic thinking is challenged by the rise of China and the relative decline in US military capabilities and Japan's economic influence.

---

[1] In 1997, Koichi Kato, the Secretary General of the Liberal Democratic Party (LDP) stated that Japan-US-China should be equidistant one at the Council of Foreign Relations in the United States. *"Kato Kanjicho Kyou-kikoku,"* (Secretary General Kato is back from the US today) *Yomiuri Shimbun*, July 26, 1997. In 2006, Ichiro Ozawa, the leader of the Democratic Party of Japan told Chinese President Hu Jingtao that Japan-US-China should be equidistant one in China. *"Nichi-bei-chu Seisankakkei-ron Futatabi,"* (Japan-US-China equidistant idea comes back again) *Sankei Shimbun*, July 5, 2006.

[2] Eric heginbotham and Richard J. Samuels, "Japan's dual hedge," *Foreign Affairs*, September/October 2002.

I . Japan and China in 1970s and 1980s: The "Friendship Paradigm" and the Golden Age of Good Relations

Japan regained independence in 1952 when the San Francisco Peace Treaty with Allied powers came into force. However, the People's Republic of China and the Soviet Union did not sign the treaty. And while Japan signed a separate peace treaty with the Republic of China in Taiwan, as recommended by the United States, Japan did not restore diplomatic relation with the PRC until 1972.

This was just after the breakthrough visit by US President Richard Nixon to the PRC in 1972. Prime Minister Kakuei Tanaka and Foreign Minister Masayoshi Ohira paved the way in normalizing Tokyo's ties with Beijing, although they faced internal resistance from conservative, pro-Taiwan politicians within the ruling Liberal Democratic Party (LDP). The majority of the Japanese public, though, welcomed the normalization with the PRC.

There were high expectations of the PRC, especially regarding commercial interests, among Japanese political and business leaders. Beijing's skillful diplomacy contributed to Japanese good feelings toward China, such as Chairman Mao Zedong's generous expression of forgiveness for Japan's past aggression and Premier Zhou Enlai's skilful renunciation of reparation demands.[3]

The result was a Japan-China "friendship" paradigm defined by good will on both sides. A more realism-based interpretation of the "friendship" would be that Japan and China, along with the United States, shared a strategic interest in containing the Soviet Union as a common threat and that Japan had commercial interests in mainland China's potentially enormous market. It is important to point out that feelings of remorse over war aggressions in China were shared across the Japanese political spectrum, from the

---

[3] Tatsumi Okabe, "Nit-Chu kankei no kako to shorai," (The Past and Future of Sino-Japanese Relations) *Gaiko Forum*, February 2001, pp. 12–13.

leftists in the Japan Socialist Party (JSP) to the rightists in the LDP, who participated in or witnessed Japan's aggressions against China in the 1930s.

The friendship paradigm continued to work and resulted in actual mutual cooperation in the 1980s after the peace treaty between Japan and the PRC was signed and came into force in 1978. Kazuko Mori points out that there were four background factors to the friendly bilateral relations in the 1980s. First, China initiated its "reform and opening up" policy, led by Deng Xiaoping. Second, US-China relations were good following normalization in January 1979. Third, the economies of Japan, the United States, and the Asia-Pacific region were growing. And fourth, Japanese Prime Minister Masayoshi Ohira launched a new face phase economic cooperation for the Asia-Pacific region as a new strategy.[4]

Ohira, who was foreign minister during the normalization talks with China, visited China and agreed to provide government loans for urgent infrastructure projects in December 1979. This marked the start of over 3 trillion yen in concessionary loans to China through 2003. In 1978, China made a bold, strategic decision to accept foreign assistance from capitalist countries in order to develop its economy, and Japan became the first capitalist country to become a donor of development assistance to China.

Thus, good relations between China and Japan in the 1980s were cemented with China's expectations of Japanese assistance and strategic relations with the United States against a common, strategic rival—the Soviet Union. Japan's aid policy toward China in the 1980s was a reflection of the strategic notion that helping the modernization of the Chinese economy and the leadership of Deng Xiaoping and Hu Yaobang would serve the interests of both the Western capitalist bloc and Japan.[5] Japan-China strategic coop-

---

[4] Kazuko Mori, *Nit-Chu kankei* (Japan-China Relations) (Tokyo: Iwanami Shinsho, 2006), p. 109.

[5] *Ibid.*, p. 109.

eration within the friendship paradigm continued until around 1990, when it came to be challenged by drastic structural changes following the collapse of the Soviet Union in 1989.

Ⅱ. Domestic Political Dynamics behind Japan's Constitutional Pacifism

Under the friendship paradigm, Japan's national security policy struck a balance between the country's alliance with the United States and self-restraint in an effort not to provoke its neighbors, which had experienced Japan's aggressions or colonial policy. This was possible because China's primary security concern was the Soviet Union, rather than Japan, although China was not fully comfortable with Japan's rearmament, which started in 1950 with the outbreak of the Korean War through the establishment of the prototype of the current Self-Defense Forces.

In the 1970s and 1980s, Japan's defense policy was seriously constrained by the political rivalry between the ruling Liberal Democratic Party (LDP) and the opposition Japan Socialist Party (JSP). The LDP favored maintaining a close alliance with the United States and securing business opportunities under a free market economy, whereas the JSP was sympathetic toward the Communist bloc and maintained close ties with the Soviet Union and the PRC. The JSP asserted its pacifist role in preventing the conservative LDP from returning to prewar militarism, which devastated Japan's Asian neighbors, by appealing to the antiwar sentiments of the people, many of whom still carried vivid memories of their own suffering in World War II. The JSP strategy was to utilize the so-called war-renouncing clause of Article 9 in the Constitution to limit Japan's full rearmament and military cooperation with the United States, notably by interpreting the Constitution as banning collective security arrangements.

Michael Green describes role of the JSP as "Greek chorus for pacifism" and "the Constitution discouraged LDP from pursuing divisive foreign and security policy initiatives". He explains JSP's tactics as a leverage to "hold up otherwise popular LDP budgets in the Diet."[6]

Yet, such a political constraint on Japan's defense policy was not a critical impediment to its alliance with the United States in the 1970s and 1980s during the Cold War, when any potential contingencies involving Japan were premised on military engagements against the Soviet Union. Such contingencies could effectively be deterred through "mutually assured destruction" arrangements between the United States and the Soviet Union. In addition, US-China strategic cooperation against the Soviet Union contributed to mitigating regional conflicts around Japan. As a result, Japan could concentrate on its own territorial defense through close military cooperation with the United States. In this context, Japan needed only to exercise its right of individual self-defense, rather than collective self-defense.

By its nature, the LDP's policy stance seeking a closer security alliance with the United States was in conflict with the JSP's ideological sympathy toward the Chinese Communist Party. But such policy differences did not come to a head within the friendship paradigm, mainly because LDP and JSP leaders shared a feeling of remorse toward Japan's aggressions in China in the 1930s.[7] For the LDP government, the US strategic shift to cooperate with the PRC against the Soviet Union was the last push needed to normalize ties in 1972.

---

[6] Michael J. Green, *Japan's Reluctant Realism: Foreign Policy Challenges in an Era of Uncertain Power,* (New York: Palgrave, 2003), p.38.

[7] There were two factions in the JSP, one pro-China and the other pro-Soviet, and the two were occasionally at odds. Tomomi Narita, who was chairman from 1968 to 1977, belonged to the pro-China faction and contributed to the secret negotiations with the PRC before normalization in 1972. He also led the drafting of the joint communiqué issued by the JSP and the Japan-China Friendship Association in 1975 that included an anti-Soviet hegemony clause.

Under the friendship paradigm, the Japanese government maintained a large development assistance program for China. Throughout the 1970s and 1980s, the Japanese people had a friendly perception of China, with public opinion polls showing that around 70% had a favorable view of China.[8]

III. Paradigm Shift toward "Normalcy" and "Realism" in the 1990s

Favorable feelings toward China declined sharply following the 1989 Tiananmen Square incident, in which the Beijing government violently oppressed a pro-democracy movement. Still, the Japanese government was one of the first to resume relations with China after economic sanctions were imposed by Western democracies to protest human rights violations. In the 1990s, commercial interests played a major role in Japan's continuation of economic assistance, even after the end of the friendship paradigm.[9]

As the Chinese economy grew, anxiety and frustration over the Chinese attitude increased among the Japanese, such as when Chinese leaders chose to lecture the Japanese over history issues. When Chinese President Jiang Zemin visited Japan in December 1998, his strident demands for an apology for past aggressions disappointed even those Japanese who believed in the importance of Japan-China relations. This was a reflection of fatigue among the Japanese people after repeated expressions of remorse over the history issues. Jiang did not express any gratitude for Japan's recent 390 billion yen economic assistance to China, and many Japanese people started to feel that China would never stop criticizing Japan, no matter how much assistance was provided as long as the "history card" worked in China's interests. During President Jiang's visit to Japan, Japanese Prime Minister Keizo Obuchi

---

[8] Public opinion poll conducted by the Cabinet Office of Japan, <http://www8.cao.go.jp/survey/y-index.html#nendobetsu>.

[9] Michael Green and Benjamin Self, "Japan's Changing China Policy: From Commercial Liberalism to Reluctant Realism," *Survival*, Vol. 38, No. 2 (Summer 1996).

refused to include an apology in the joint communiqué, although he was willing to do so following a summit with South Korean President Kim Dae-jung.

The Japanese frustration was partly due to the prolonged economic slump after the collapse of the economic bubble in 1989, while China was enjoying an economic boom. The Japanese were also annoyed that China actively persuaded many Asian and African countries to vote against a resolution to expand the number of permanent members in the UN Security Council, submitted by Japan, Germany, India, and Brazil in July 2005. Many of those Asian and African nations voting against the resolution had received financial aid from China. This was quite shocking to the Japanese, as Japan had extended a total of around 6 trillion yen in development assistance to China between 1979 and 2007.

This was a period, though, when frustrations and anxieties regarding China were more emotional than based on a perception of actual economic or national security threats. The size of the Chinese economy and the modernization of the Chinese military were not yet serious concerns for the majority of the Japanese public.

IV. China's Growing Military in the 1990s

Defense experts and officials first acknowledged that the military posture and actions of China posed a potential future threat to Japan's security in the 1990s. In 1994 General (retired) Ikuo Kayahara, a defense expert on China's military at the National Institute of the Defense Studies, examined Japanese anxiety over the potential military threat that China's military modernization represented. He pointed to four concerns: (1) the large size of the military, (2) the tendency of military budgets to increase, (3) attempts at territorial expansion using military strength, and (4) the modernization of China's nuclear arsenal and its proliferation to other countries.

China had the world's largest armed forces (3.03 million troops), the third largest naval fleet as measured by tons, and the second largest number of military airplanes. Thus the sheer size of the armed forces was felt to be a threat for many Japanese nationals.

Secondly, Kayahara pointed out that China's military budget had been registering double-digit growth since 1989. Anxiety in Japan was reinforced by the opaque nature of the allocations, such as the omission of procurement costs for expensive military equipment from the total military budget.

Thirdly, the unilateral claim to sovereignty over the Spratly Islands with the enactment of the Law of the Territorial Sea and Contiguous Zone in 1992 stirred Japanese suspicions, given China's expansionary past.

Fourthly, China's nuclear test on October 5, 1993, despite pressure from the United States and the modernization of its nuclear weapons, such as warheads for intermediate-range ballistic missile, which can target its Asian neighbors, caused alarm in Japan in the midst of moves toward nuclear disarmament between the United States and Russia. Kayahara noted that China's position on the transfer of both nuclear and conventional weapons to such countries as Pakistan and the failure to dissuade Pyongyang from pursuing nuclear development were additional sources of Japanese suspicions. Kayahara concluded that public concerns about China were largely legitimate.[10]

In 1996, China conducted a missile launch exercise as Taiwan was about to hold its first democratic presidential election. This prompted Japan to strengthen its alliance with the United States. In 1997, Japan and the United States agreed to review the Guidelines for

---

[10] Ikuo Kayahara, *Chugoku gunjiron* (On Chinese Military Affairs) (Tokyo: Ashi Shobo, 1994), pp. 34-43.

U.S.-Japan Defense Cooperation, broadening the scope of bilateral security cooperation to areas surrounding Japan, which was not covered in the previous guidelines of 1978.[11]

V. China's Reaction to the 1997 Security Guidelines

The 1997 guidelines enabled Japan to provide logistic support for the US military in the event of a contingency in areas surrounding Japan. Yet, it still carefully avoided Japan's direct support for US combat activities, which could be regarded as a violation of the constitutional ban—as interpreted by the government—on the exercise of the right of collective self-defense. Presumed Japan-US contingency cooperation was for noncombat activities, such as relief for refugees, search and rescue, noncombat evacuation operations, or the inspection of ships in support of UN economic sanctions.[12] The Japanese government continued to interpret Article 9 of the Constitution as not allowing Japan to exercise its right of collective self-defense.

The "contingency" assumed in the 1997 Guidelines was one involving the Korean Peninsula, as North Korea had suggested it might target US bases or neighboring countries while pursuing the development of nuclear arms. However, the guidelines could theoretically also be applied to a contingency in the Taiwan Strait, and this worried China.

Koichi Kato, a pro-China liberal Diet member and LDP secretary general at the time visited China and explained to Chinese officials that the guidelines were not targeted at China but at North Korea. Immediately after Kato's visit, though, Chief Cabinet Secretary Seiroku Kajiyama commented

---

[11] Japan Defense Agency, *Defense of Japan* 1997, pp. 168-69.

[12] Ibid., pp. 331-32.

that the Taiwan Strait would not be excluded. Japan thus wound up sending a mixed message to China.[13]

There was another incentive behind the 1997 guidelines. It was to reestablish the trust in the Japan-US alliance that was damaged following the rape of a Japanese elementary schoolgirl by US military personnel in Okinawa.

Whatever the motivation, Japan's attempt to improve its security cooperation with the United States ended up provoking China in the 1990s. China's reaction became harsh and negative. At the same time, Kato's attitude and the lack of public criticism over his remarks suggest that the Japanese public did not regard China as a serious security threat in the 1990s, although military experts like Kayahara expressed substantial concerns about China's military modernization. The public was more critical of Kato's conciliatory stance some 10 years later. In 2006, the house of Kato's mother was set on fire by a right-wing activist angry at his criticism of Prime Minister Jun'ichiro Koizumi's visit to Yasukuni Shrine.[14] China was critical of visits by senior Japanese officials and warned of a resurgence in Japanese militarism.

The Japanese public became worried about China's expansionist behavior and military modernization and was fed up with repeated criticism of Japan's historical perception. Indeed, Chinese "research vessels" and warships frequently intruded into Japan's exclusive economic zone (EEZ) in the East China Sea around the Senkaku Islands in 1990. Yet, the Japanese public did not imagine that China would pose a serious threat to Japan's security. The Japanese still underestimated the impact of China's rapid economic

---

[13] Nicholas D. Kristof, "For Japan, a Quandary on Pleasing Two Giants," *The New York Times,* August 24, 1997.

[14] "Home grown political terror," *Japan Times,* August 20, 2006.

growth and its firm determination to challenge Japan's territorial claim over the Senkakus and the surrounding EEZ.

VI. Shaky Japanese Political Leaderships

During his tenure as prime minister from 2001 to 2006, Jun'ichiro Koizumi was determined to visit Yasukuni Shrine, where Japan's war dead, including class-A war criminals, are enshrined.

Security dialogue between Japan and China, which had been agreed upon in the bilateral summit of 1998, was frozen due to the Yasukuni controversy. It was not resumed until five years later, when Defense Minister Shigeru Ishiba visited China in 2003.

Political relations with China became cold, but economic ties were still quite robust. Japan's exports to China increased from $23.3 billion in 1999 to $80.34 billion in 2005.[15] Japan's annual direct investment in China increased from $360 million in 1999 to $6.58 billion in 2005.[16] Such economic interdependence was a source for optimism about China's future trajectory, despite the cold reality of mutual suspicion and frustration.

The National Defense Program Guidelines (NDPG), FY2004 and beyond of the Japanese government showed restraint and did not call China a threat. Among the major threatening factors listed were such new threats and situations as the proliferation of weapons of mass destruction, ballistic missiles, and international terrorism. It added that the probability of a full-scale

---

[15] "Nicchu bouekigaku no suii," (Past annual Japan-China trade amount) *Japan's Ministry of Finance Website*, <http://www.mofa.go.jp/mofaj/area/china/boeki.html>.

[16] *"Nihon no kuni chiiki-betsu taigai chokusetsu toshi,"* (Japan's foreign direct investment to countries) *Japan External Trade Organization (JETRO) Website*, <https://www.jetro.go.jp/world/japan/stats/fdi/>.

invasion against Japan had declined. NDPG FY2004 only noted that Japan should be attentive to future actions by China.[17]

In 2006, Prime Minister Shinzo Abe, who succeeded Koizumi, showed his desire to repair Japan's relationship with China by visiting the country as an "icebreaking" trip within two weeks of becoming prime minister in October 2006. He showed restraint in his approach to the history issue, despite his conservative views. Japan and China agreed to seek an acceptable resolution of issues related to the East China Sea and to resume bilateral security dialogue and defense exchange. In return, Chinese Premier Wen Jiabao made an "ice-melting" trip to Japan in April 2007.[18] Political relations between the two countries seemed to resume the smooth course of earlier years.

However, this was the start of a period of domestic political turmoil in Japan, with prime ministers changing every year. Each administration had its own diplomatic stance, and such incessant change in the political leadership eroded the mutual trust that had been nurtured until then.

China's expectations regarding Japan continued under Prime Minister Yasuo Fukuda, who succeeded Abe one year later after Abe stepped down in September 2007 due to illness. Fukuda showed a willingness to improve bilateral relations, and China responded positively to Fukuda's show of goodwill. It proposed the joint development of undersea gas fields in the East China Sea, and the two countries agreed on a joint development plan in June 2008. Fukuda, though, resigned suddenly in August 2008 in the face of a political gridlock.

---

[17] Cabinet Public Relations Office, *Heisei 17 nendo iko ni kakaru boueikeikaku-no taiko ni-tsuite* (National Defense Program Guideline for FT 2004 and beyond), 2004, <http://www.kantei.go.jp/jp/kakugikettei/2004/1210taikou.html>.

[18] Speech by Premier Wen Jiabao of the State Council of the People's Republic of China at the Japanese Diet, Tokyo: April 12, 2007. English translation is carried at the website of the Consulate-general of the PRC at Manchester, <http://manchester.china-consulate.org/eng/xwdt/t311107.htm>.

He was succeeded by Taro Aso, whose foreign policy plan known as the "arc of freedom and prosperity" surprised China, as it appeared to promote the country's encirclement by democratic nations. China retreated from the joint development agreement and starting drilling in the East China Sea without first consulting Japan.

Seeds of mistrust continued to be sown. In the general election of August 2009, the Democratic Party of Japan defeated the LDP and created a non-LDP government for the first time since 1955 (except briefly in the early 1990s, when the LDP was still the largest party in the Diet). Due to his immature diplomatic views and lack of experience, though, DPJ Prime Minister Yukio Hatoyama only created confusion in Japan's foreign policy. His wavering stance on the relocation of the Futenma base in Okinawa seriously damaged the Japan-US alliance. Hatoyama was keen to improve relations with China and South Korea, though, and stated Japan would seek to ease its heavy dependence on the United States. During the bilateral meeting with Premier Wen in June 2010, he accepted China's demand to downgrade the existing agreement on undersea gas fields in the East China Sea from joint development to Japan's partial investment.

VII. Collision with Chinese Fishing Boat and Japanese Government Purchase of the Senkaku Islands

Due to his mishandling of the Futenma issue, Hatoyama was forced to resign and Naoto Kan became prime minister in September 2010. One of Kan's priorities was to repair the damaged Japan-US alliance. Almost immediately after Kan entered office, an accident occurred in which a Chinese fishing boat collided into Japan Coast Guard patrol boats near the Senkaku Islands. The inexperienced administration's handling of the situation surprised and provoked China, as the boat captain was detained and became subject to criminal prosecution.

This surprised the Chinese government, as the Japanese practice until then had been to "catch and return" any Chinese "intruders" without taking any legal action. "Catch and return" was a gentleman's agreement to avoid complicated legal procedures, which could lead to questions of sovereignty. Incentives for the Kan administration to change the rules of the game are still not clear. It may have been a combination of inexperience and a lack of informal communication channels with China, which LDP administrations had maintained. Unfortunately, the minister supervising the Japan Coast Guard at the time was Seiji Maehara, who had been criticized as an anti-China nationalist after stating that China was a threat to Japan in a speech to a US think tank in 2005.[19] China may have been suspicious of an emerging nationalistic streak with the change in Japan's game rules. Since the collision occurred during a transition period, Maehara remained in the cabinet as Foreign Minister to negotiate with China. Kan was not deliberately choosing an anti-Chinese minister but needed Maehara—who had restored trust with US security experts as foreign minister—to repair the Japan-US alliance that was damaged during Hatoyama administration. This choice, though, may have offended China.

China's tough reaction to the Japanese move also surprised the Japanese government. China demanded the immediate release of the detained captain, arrested four Japanese businessmen stationed in China on suspicion of espionage activities and restricted exports of rare earth metals—essential for the production high-tech electric devices by Japanese manufacturers.[20]

China's use of economic tools to achieve diplomatic ends made the Japanese very nervous. Before it happened, the Japanese never imagined that China would take such a harsh measure, which could wind up hurting Chi-

---

[19] "China's Peaceful Development Poses No Threat," *China Daily,* December 14, 2005, <http://www.china.org.cn/english/2005/Dec/151840.htm>.

[20] Martin Fackler and Ian Johnson, "Japan retreats with release of Chinese boat captain," *The New York Times,* September 24, 2010.

na's own business interests and reputation as well. Even during the height of the Yasukuni controversy in the Koizumi years, political differences did not spill over into economic and business affairs.

Anxiety over Chinese retaliatory tactics were further aggravated in September 2012 as large-scale anti-Japanese demonstrations turned violent, leading to the destruction of Japanese cars, factories, and stores. The demonstrations were triggered by the Japanese government purchase of three islands in the Senkaku chain from a Japanese private owner. China had warned the Japanese government not to make the purchase, since this could be regarded as an exercise of sovereignty. The timing could not have been worse. It was a very sensitive moment in the power transition from President Hu Jintao to Vice President Xi Jinping in preparation for the National People's Congress in November.[21] Many observers pointed out that relations between Japan and China reached their lowest point since ties were normalized in 1972.

VIII. Second Abe Administration

The LDP, led by former Prime Minister Shinzo Abe, defeated the ruling DPJ in the December 2012 general election, and returned to power. One major reason for the victory was people's frustrations with the DPJ's immature handling of economic and foreign policy since the Hatoyama administration in 2009. People were especially critical of Hatoyama's mishandling of the US-Japan alliance and the relocation of the Futenma base in Okinawa. Peo-

---

[21] China expert, Ryosei Kokubun, President of the National Defense Academy points out that internal rivalry in Chinese leadership complicated Senkaku issues. "Nicchuu mondai o kangaeru," (Consider Japan-China relation) *Nikkei Shimbun*, September 28, 2012.

ple generally believe the DPJ administrations damaged both the Japan-US alliance and relations with China.[22]

Abe was not considered a particularly popular figure, though, as many people remembered his miserable first term as prime minister from September 2006 to September 2007. The LDP's landslide victory in December 2012 can thus be seen as being more of an anti-DPJ vote than an expression of hope in the LDP; the public was placing its expectations on the LDP's half a century of experience in managing government, rather than on Abe's personal ability.

During the first year of his second term, Abe consistently received high approval ratings, for he succeeded in revitalizing a stagnant economy with an effective monetary policy and massive economic stimulus spending, a policy dubbed "Abenomics."

Abe's strong commitment to the Japan-US alliance and his realistic foreign and security policy, as reflected in the administration's revision of the National Defense Program Guidelines for FY 2014 and beyond, the founding of the National Security Council (NSC), and the issuance of Japan's first National Security Strategy (NSS) in December 2013, have gained the support of the Japanese security community.

NDPG for FY2014 and beyond contains a concept called Dynamic Joint Defense Force, which represents an attempt to enhance Japan's defense mobility and the capability to defend its own territory, especially around the Nansei Islands southwest of mainland Japan, where China has increased its military and paramilitary activities. The NDPG also seeks greater security cooperation with like-minded countries in the Asia-Pacific region, as well as

---

[22] Leonard Schoppa, "A Vote against the DPJ, Not in Favor of the LDP," *Newsletter: Japan Chair Platform,* December 18, 2012, *Center for Strategic & International Studies website,* <http://csis.org/publication/vote-against-dpj-not-favor-ldp>.

with the United States.[23] The NSS clearly defines the rationale behind and sets the direction for Japan's security strategy in an open document. The NSS also proposed that Japan proactively cooperate with Asian neighbors, such as by providing coast guard ships as a form of capacity building assistance.[24]

While the Japanese public welcomes Abe's economic policy and realistic alliance policy, people are worried about the continued chilly relations with China and South Korea. This is a negative inheritance of the previous Yoshihiko Noda administration, during which time Japan became embroiled in territorial issues with China (Senkaku) and South Korea (Takeshima). One additional bone of contention that has emerged is Abe's perceptions of history and the reported attempts to revise past Japanese government statements apologizing for Japan's past aggressions and recruitment of "comfort women" from the Korean Peninsula during World War II.

The Chinese and South Korean governments severely criticized Prime Minister Abe's December 2013 visit to Yasukuni Shrine. The visit also elicited an unusual statement expressing "disappointment" from the US government, which was worried that the move could escalate regional tensions, possibly setting off an accidental conflict around the Senkaku Islands or pushing South Korea, another US ally, toward China.[25]

Washington's worries about Abe's perceptions of history appear legitimate, as his security and foreign policy has intentionally or unintentionally

---

[23] Japan Ministry of Defense, *National Defense Program Guidelines for FY2014 and Beyond* (provisional translation), December 17, 2013, <http://www.mod.go.jp/j/approach/agenda/guideline/2014/pdf/20131217_e2.pdf>.

[24] Japan Ministry of Defense, *National Security Strategy (provisional translation)* December 17, 2013, <http://www.mod.go.jp/j/approach/agenda/guideline/pdf/security_strategy_e.pdf>.

[25] Hiroko Tabuchi, "With shrine visit, leader asserts Japan's track from pacifism," *The New York Times*, December 26, 2013.

frequently been misunderstood or exaggerated. Heightening tensions could undermine the legitimacy of the Japan-US alliance and the stability of the Asia-Pacific region.

In its December 18 editorial, for example, the *China Daily* warned against Abe's "proactive pacifism," asserting that "the catchy but vague expression" is "Abe's camouflage to woo international understanding of Japan's move to become a military power."[26]

Abe's intentions, however, are not to turn Japan into a military power, either in qualitative or quantitative terms. Rather, his security policy is designed to incrementally enhance the functionality of Japanese defense capacity.

The *China Daily*'s editorial pointed out that Abe's doctrine seeks to turn Japan's Self-Defense Forces into "ordinary armed forces." In reality, though, the SDF are far from "ordinary armed forces," which are able to take necessary actions to defend the country and to contribute to regional security promotion. Japan's defense-related legal system is so restrictive, in fact, that the inability to respond to contingencies—even when a defense response is required—could actually abet the escalation of tensions with neighbors.

On February 4, 2014, the prime minister hosted a meeting of an expert panel on reconstructing the legal basis for national security. The panel proposed legislation to enable the use of the SDF to deal with so-called "gray zone" situations categorized between peacetime and genuine contingencies.[27] Under current Japanese law, the country cannot exercise its right of self-defense unless it is under organized, armed attack.

---

[26] "Abe's three shots at pacifism," *China Daily*, December 18, 2013, <http://www.chinadailyasia.com/opinion/2013-12/18/content_15106729.html>.

[27] "Anzenhosho no houteki kiban no saikouchi ni kansuru kondankai," (Advisory Panel on Reconstruction of the Legal Basis for Security) *Japanese Prime Minister's Office website*, February 4, 2014, <http://www.kantei.go.jp/jp/96_abe/actions/201402/04anpo.html>.

The recommendation to the prime minister came from a panel of cool-headed defense experts, not emotional nationalists. The Tokyo Foundation, an independent nonprofit think tank, made above point in its policy proposal on "Maritime Security and the Right of Self-Defense in Peacetime," released in November 2013 under a project for which I myself served as leader. The recommendations of Abe's panel is closely aligned with the Tokyo Foundation's proposal, which called on the government to make reforms to the Diet, law enforcement authorities, and the SDF that are long overdue in order to more effectively defend Japan's territory and avoid an escalation of tensions with its neighbors, especially China, which is now persistently sending paramilitary vessels into Japan's territorial waters.[28]

Currently, the deployment of the SDF is heavily restricted by legal and political concerns, even in addressing self-defense needs. This is based on Japan's remorse for the suffering caused to its neighbors, including China, by its wartime aggressions. This self-restraint was functional during the Cold War, since Japan's exercise of self-defense was chiefly directed against the Soviet Union and integrated into US military strategy. There was no need to address intrusions into its territorial waters by paramilitary vessels.

A military invasion of Japan would be a clear case of a contingency, when the Japanese government can legally order the SDF into action. Considering the current situation surrounding the Senkaku Islands, though, Japan is more likely to face minor yet critical challenges from nonmilitary or paramilitary vessels, which would not be considered armed attacks. This could place the Japanese government in a dilemma. The SDF cannot use their full military capabilities without a defense mobilization order from the govern-

---

[28] The Tokyo Foundation, *Maritime Security and the Right of Self-Defense in Peacetime: Proposals for a National Security Strategy and the New National Defense Program Guidelines (translation of the summary)*, May 12, 2014, <http://www.tokyofoundation.org/en/articles/2014/maritime-security-and-self-defense-in-peacetime> .

ment clearing the way for self-defense maneuvers. If the government does issue such an order for an incident around the Senkakus, though, this could send the wrong signal.

Ordinary democracies, such as the United States and its European allies, do not have such a dilemma, since predefined Rules of Engagement outline the actions to be taken by their military forces. Japan is constricted by its deep remorse for past military aggressions and understands the sensitivity of its neighbors. But we must also keep in mind that lapses in Japan's national security laws could actually lead to a heightening of tensions in the East China Sea.

Even if the legal reforms are legitimate, many criticized why the prime minister needed to visit Yasukuni Shrine, stirring up new controversy and worrying neighbors. I agree that the visit was ill-timed, but we live in an imperfect world in which emotional nationalism can sometime become a source of political capital. This is true not only in Japan, though, but also in China, South Korea, and even the United States, which saw an upsurge of patriotic sentiment following the September 11 terrorist attacks.

Despite the image encouraged by China and South Korea, the Abe administration is marked more by realism than nationalism. University of Tokyo Professor Emeritus Shin'ichi Kitaoka, who is deputy chairman of Abe's panel on reconstructing the legal basis for national security, was the leader of a Tokyo Foundation project on Redefining Japan's Global Strategy, which also recently announced its policy proposal. The core message of the proposal was the importance of restraining emotionalism and taking pragmatic steps to find common ground with its neighbors.[29]

---

[29] The Tokyo Foundation, "Abe gaiko-eno 15 no shiten-Nashonarisumu yori riarizumu no tsuikyu o," (Pragmatism over Nationalism: 15 Foreign and Security Policy Choices for Prime Minister Abe) August 2013, <http://www.tkfd.or.jp/files/doc/2013-02.pdf>.

The prime minister's visit to Yasukuni Shrine was not fully supported by Japanese general public and realistic strategic thinkers. Three (Yomiuri, Asahi, Nikkei) out of four major news papers' editorials criticized Prime Minister Abe's visit to Yasukui Shrine. Centrist Nikkei Newspaper poll shows that 45% supports Abe's visit while 43% opposes in January 2014.[30] The administration's current security policy initiatives, therefore, are not the result of an emotionally charged nationalism but represent a rational and incremental development of democratic governance in Japan's postwar security and defense policy.

IX. Four Strategic Choices regarding the Rise of China

In an essay published in 2007, Mike Mochizuki introduced four patterns in Japan's strategic options toward China: cooperative engagement with a soft hedge, competitive engagement with a hard hedge, balancing and containment, and strategic accommodation.[31]

Mochizuki pointed out in the essay that a majority of Japanese foreign policy experts advocate cooperative engagement with China, noting that China desires a peaceful international environment in order to pursue wealth and power. This school, he says, acknowledges that China's development may engender new international challenges, such as opaque military modernization, hyper energy demand, or spillover from socioeconomic turmoil due to domestic contradictions, which should be met by strengthening Japan's military alliance with the United States and engaging in dialogue with China. Mochizuki explained that the school is distinct from "the cooperative school with soft hedge" since those in the former "shy away from explicitly

---

[30] "Shusho no yasukuni sannpai sanpi wakareru," (Prime minister's Yasukuni visit splits public opinions) *Nikkei Denshiban*, January 25, 2014, <http://www.nikkei.com/article/DGXNASFS26023_W4A120C1PE8000/>.

[31] Mike M. Mochizuki, "Japan's Shifting Strategy toward the Rise of China," *The Journal of Strategic Studies*, August-October 2007.

characterizing the US-Japan alliance as a tool for balancing or containing China."[32]

Mochizuki saw "the competitive engagement with hard hedge" school as being more skeptical of the possibility that China will adhere to a cooperative strategy as its power capabilities increase. This school is also worried that the balance in conventional weapons could tip in favor of the Chinese side, even in comparison to the combined air and naval capabilities of Japan and the United States. Mochizuki chose "competitive" engagement rather than "cooperative," since the school sees Chinese leaders using the "history card" to "weaken Japan's to resolve to stand up to China and to reduce its regional influence."[33]

Mochizuki distinguished "balancing and containment" from "competitive engagement with hard hedge," saying that the former camp believes that China will eventually embrace hegemonic ambitions based on cultural or historical reasons and could, in the future, face internal turmoil due to socio-political contradictions. Some members of the school raised doubts about the credibility of US extended deterrence in the context of China and proposed that Japan pursue its own nuclear option.[34]

Mochizuki pointed out that the "strategic accommodation" school differs from "cooperative engagement with soft hedge" as it believes that tightening the alliance with the United States could jeopardize cooperative ties with China.[35]

---

[32] *Ibid.*, pp. 759-762.

[33] *Ibid.*, p. 762.

[34] *Ibid.*, p. 765.

[35] *Ibid.*, p. 766.

X. China as a Security Threat

The Japanese perception of China has deteriorated through the chain of events that created mutual distrust in the 2000s and early 2010s. Although Japan occasionally experienced diplomatic skirmishes with China in the 1980s and 1990s, the relationship had the ability to get over them. Japan's experience with China in the 2000s, however, has prompted many to regard Chinese as a potential threat to Japan's survival.

Public opinion polls conducted by Japan's Cabinet Office shows the changes in Japanese perceptions of its giant neighbor. In 2000, the biggest threats to Japan's national security were regarded as being the Korean Peninsula (56.7%), disarmament and weapon of mass destruction (35.2%), and US-Russia relations (17.9%). China was not even among the choices offered except in the context of US-China relations (13.1%) and China-Russia relations (11.7%). The Japanese did not assume that China itself was a security threat to Japan.[36]

In the same poll conducted in 2012, many Japanese respondents said they were concerned about the modernization of China's military and its maritime activities (46.0%)—the second most popular response after the Korean Peninsula (64.9%).[37]

Another poll conducted by a Japanese NGO in 2013 showed the Japanese feel most threatened militarily by North Korea (73.4%), followed by China (61.8%). Respondents choosing China also gave their reasons why. The two top choices were concern with China's intrusions into Japan's terri-

---

[36] Japan's Cabinet Office, "Jieitai bouei mondai ni kansuru yoronchosa," (Public poll regarding to the Self-Defense Forces and defense issues) January 2000, <http://www8.cao.go.jp/survey/h11/bouei/>.

[37] Japan's Cabinet Office, "Jieitai bouei mondai ni kansuru yoronchosa," (Public poll regarding to the Self-Defense Forces and defense issues) January 2012, <http://www8.cao.go.jp/survey/h23/h23-bouei/index.html>.

torial waters, and concern with Japan and China's conflict over territories, which were more 60%. The same poll asked the same questions to Chinese citizens, who saw the United State as being the biggest threat (71.6%) and Japan as the second biggest (53.9%).[38]

The poll results suggest a growing perception of China as a threat during the 2000s. The first turning point was Prime Minister Koizumi's Yasukuni visit in 2005. According to the same poll in 2006, 42.8% of Japanese respondents said they see China as a military threat.[39] Yet, the Japanese were optimistic that mutual economic interdependence would prevent military conflict.

The second turning point was 2010 after the fishing boat collision near the Senkaku Islands. The Japanese realized that China would resort to economic measures to address diplomatic issues even if they may damage their own economic interests. It may not be a coincidence that it was in 2010 that China's gross domestic product surpassed Japan's to become the second largest in the world. The Chinese economy was now big enough to allow for a few dents in economic relations with Japan. This is a worrisome trend for the Japanese, as it suggests there could be more retaliatory actions on the economic and business front, as evidenced by China's harsh reaction to the Japanese government's purchase of the Senkaku Islands in September 2012.

XI. Historical Limitation and the Japan-US-China Triangle

As noted above, China was not seen as a serious security threat until very recently. China, though, has seen the Japan-US military alliance as being against their security interests. Historically and structurally, the United

---

[38] Joint Poll by Genron NPO and China Daily, conducted in May to July 2013, *Genron NPO Website*, <http://www.genron-npo.net/pdf/2013forum.pdf>.

[39] Joint Poll by Genron NPO and China Daily, conducted in June to July 2006, *Genron NPO Website*, <http://tokyo-beijingforum.net/index.php/survey/2nd-survey>

States has been an integral part of Japan's security strategy toward China. In 1930s and 1940s, Imperial Japan invaded China with colonial ambitions. It was the United States, which provided military support to Chiang Kai-shek, the leader of the Republic of China on the mainland against Japan's invasion. In 1941, Japan attacked Pearl Harbor to start a war against the United States. After its defeat in World War II, Japan's rearmament was driven by the US strategy for the emerging Cold War. Japan recovered its independence and began limited rearmament under a new Constitution that included a clause vowing to "renounce war as a sovereign right of the nation and the threat or use of force as means of settling international disputes." This idealistic clause was augmented with the US military presence in Japan under a mutual security treaty to guarantee Japan's security. For China, the Japan-US security arrangements were occasionally seen as a threat to its own security. China was more antagonistic toward the United States in the early stages of the Cold War and during Korean War than at any other time. China rewrote its former positive view of the United States in more antagonistic terms during this era. At the same time, the US military presence in Japan and the constitutional restraints on an independent defense policy were designed to prevent the reemergence of Japanese militarism. This facilitated China's decision to normalize ties with the US and Japan after 1972 in the face of their urgent need to counter the Soviet Union.

At the same time, Japan's economy grew more dependent on the markets of neighboring Asian countries, especially in Southeast Asia. This was part of a US Cold War strategy to turn Japan into a showcase of the success liberal democracies can enjoy and to encourage it to become a reliable security partner, guaranteeing a US military presence in East Asia against the Soviet Union and the PRC. After the US strategic shift to seek cooperation with China against the Soviet Union, China also became a very important economic partner for Japan. Strategic cooperation between China and the Japan-US alliance successfully pushed the Soviet Union toward collapse.

The side effect of the Cold War victory was that China enjoyed military and economic assistance from the United States and Japan and was able to create the foundations for its current economic success as the world's second-largest economy. As the current US strategy toward China no longer ignores the economic interdependence with China, Japan is also heavily dependent on the Chinese economy.

Another side effect of the Cold War is that Japan failed to make a moral reconciliation with Chinese and South Korean nationals regarding its past aggressions and colonial policy, although state-to-state agreements were reached. Japan's normalization with China in 1972 was achieved thanks to the strong leadership and grand strategic ideas of Mao Zedong and Zhou Enlai. There was no clear reconciliation, though, at the social level. China continued to conduct "anti-Japanese" education, since winning independence from Imperial Japan is seen as an essential source of legitimacy for the ruling Communist Party of China.

In addition, China has effectively utilized the history card to restrain Japan's aspirations to become a "normal country" by amending Article 9 of the Constitution. It was very difficult for Japanese leaders to pursue "normal nation" rearmament in the light of the strong anti-war sentiments of those who experienced World War II. In the Diet, the Japan Socialist Party, which was sympathetic to China's Marxist ideology, roused feelings of war guilt to block Japan's return to normalcy with a constitutional amendment, which requires the approval of a two-thirds majority in the Diet.

Thus, Japan's strategic options have been shaped by historical and geopolitical limitations: a security alliance with the United States, constitutional restraints on defense and security policy, diplomatic restrictions on rearmament via China's (and South Korea's) history card, and, more recently, the heavy reliance on the Chinese economy.

XII. Current Trend toward "Hard Hedge"

The current trend in Japan' strategic policy seems to be a shift toward "competitive engagement with hard hedge" and away from "cooperative engagement with soft hedge," as coined by Mike Mochizuki. Richard Bush categorizes the Tokyo Foundation's policy proposal "Japan's Security Strategy toward China" in 2011 as an example of "competitive engagement with hard hedge" school. Bush sates that the proposal focuses on strengthening Japan's own military capabilities, changing the Constitution to allow more flexibility on security policy, broadening the scope of the alliance for regional and global challenge, and expanding security ties with other like-minded nations.[40] He also added that this approach is similar to the series of policy recommendations on the Japan-US alliance contained in the Armitage-Nye Report.[41]

In Japan's two more recent administrations, led by Prime Ministers Noda and Abe, members of the "hard hedge school" have had an influential policymaking role. In fact, there is considerable continuity in the defense and security policies of the DPJ's Noda administration and the LDP's Abe administration. For example, security experts like Satoshi Morimoto and Shigeru Ishiba, whom Mochizuki categorizes as belonging to the "hard hedge" school, have played important roles in Japanese policymaking.[42] Morimoto served as minister of defense as a nonpolitical appointee in the Noda administration, and Shigeru Ishiba is the secretary general of the LDP. Morimoto's pick was regarded as an effort to reinforce the alliance with the United States, which was damaged by Prime Minister Hatoyama's unprofessional management of the Okinawa base issue. Although Ishiba was defeated by Abe in

---

[40] Richard Bush, *The Perils of Proximity: China-Japan Security Relations*, (Washington, DC: Brookings Institution, 2010), p. 269.

[41] Ibid.

[42] Mochizuki, op. cit., p.763.

the September 2012 LDP presidential election, he received more votes than Abe from non-Diet party members.(Abe won the election with higher support from Diet members.) Ishiba is known as a defense policy expert, and his support could be seen as a reflection of people's anxiety and antipathy toward China's harsh reaction to the Japanese government purchase of the Senkaku Islands in September 2012.

This apparent trend does not signal the death of "soft hedge" school, though. The small but influential New Komeito Party, which played a critical role in maintaining relations with China and continues to cooperate closely with the LDP during elections, is an influential coalition partner advocating a "soft hedge" approach. For example, the liveliest debate on the shift in security policy toward "hard hedge"—such as a reinterpretation of Article 9—are now held between the LDP and Komeito, rather than within the government or with the opposition parties.

The general trend, despite the influence of the "soft-hedge" Komeito, though, is that Japan' future trajectory will be toward more "normal country" status, both in its relationship with the United States and China. China's assertiveness in the South and Easter China Sea pushes Japan toward a closer alliance with the United States and encourages an increase in defense capacity. In the past, China's history card was effective in restraining Japan's security policy. But in the face of China's military modernization and burgeoning economy—and as the number of elder Japanese who experienced wartime aggressions declines—Japan's self-restraint is fading away.

XIII. Japan's Dilemma in the Japan-US-China Triangle

In a sense, thought, this is quite natural, as nearly 70 years have already passed since the end of World War II. The trouble is, however, historical reconciliation has not been made in the minds of the people in the PRC, who are still educated that Japan is a hostile nation. As matter of the fact, a majority of people polled in China said that Japan's current government is a military regime. This contrasts sharply with a BBC worldwide poll conducted between 2011 and 2012 that scores Japan as the nation that has had the most positive influence in the world.[43]

As Japan seeks an exit strategy from the current impasse and mutual distrust with China, it may find itself in a difficult dilemma in the trilateral relationship with the United States and China. To secure its territorial integrity, Japan needs to increase its own military capabilities and force closer alliance ties with the United States, even when such actions could raise China's anxiety or mistrust of Japan.

There is no getting around the fact that Japan needs to increase its business ties with China to secure its own economic growth. However, the Japanese are no longer so naive as to expect economic interdependence to prevent potential conflicts in the light of the Senkaku experience in 2010. The Japanese now realize that increasing economic interdependence could actually give China bigger political tools with which to limit Japan's policy choices.

Japanese strategy experts occasionally voice the concern about whether the United States would fully support Japan's security interests in the face of deepening US-China economic ties. The Japan-US-China triangle may have entered a new difficult phase as China rises into a more influential military and economic power. It behooves Japanese policymakers to address the complex nature of this triangle, in which military rivalry and cooperation, economic competition and interdependence, and historical differences and friendship are juxtaposed in the context of a dynamic power shift in the Asia-Pacific.

---

[43] BBC Poll, "Views of different country's influence," in 2011 to 2012, *Democratic Underground.com*, <http://www.democraticunderground.com/1002691517> .

Reading the Tea-Leaves of China's Force Modernisation: Capabilities, Intentions, and a Perplexed Region

Ja Ian Chong
(National University of Singapore)

Capabilities on their own, do not have to be threatening. In keeping sea lanes of communication safe from piracy and open to all seafarers, enhanced naval capabilities can provide a key regional, if not global, public good. Rapid deployment, sealift, airlift, and surveillance can support Humanitarian Assistance and Disaster Relief (HADR) as well as emergency evacuation, search and rescue, and peacekeeping. This is apparent from events like Typhoons Morakot and Haiyan in 2009 and 2013, Japan's 2012 earthquake and tsunami, the disappearance of Malaysian Airlines Flight MH370. Effective deterrence capacities can be a cause for stability. Yet, even as there is a strong desire to have Beijing play an active regional role among China's neighbours, which includes contributions based on its very significant military and paramilitary capabilities, apprehension about Chinese intentions remain.

What perplexes China's neighbours, is how Beijing intends to use its expanding military and paramilitary capabilities. Even as Beijing demonstrates great enthusiasm toward using its growing military prowess for the above types of missions, China also appears ready to use apply its newfound military capacities as coercive substitutes for regular diplomacy vis-a-vis other regional actors. This is especially apparent in Chinese approaches to maritime disputes in the East and South China Seas. Regional actors in the Asia-Pacific would like to partner Beijing in addressing both traditional and nontraditional security threats given the latter's officially stated keenness for doing so, and its capabilities. However, Beijing's readiness to use paramilitary and occasionally military forces to pressure rival claimants in maritime disputes gives regional actors pause.

Much as regional actors would like to give Beijing the benefit of the doubt and work with China, uncertainty over Chinese intentions that accompany its growing capabilities tend to prompt more cautious responses. Advanced Chinese capabilities can clearly augment anti-piracy, search and rescue, as well as humanitarian assistance and disaster relief in Asia. Regional actors generally welcome Chinese participation in the U.S. hosted Rim of the Pacific (RIMPAC) Naval Exercises in 2014, just as East Asian participants in the Cobra Gold exercises with the United States were happy to see China observe and participate in the annual event.[1] These come on top of the regular interactions with Chinese military and paramilitary forces as well as other components of its defence and security establishments that regional governments in East Asia have on bilateral and multilateral bases. Despite such regular contact and the recognition of China's ability to contribute to regional efforts, persistent uneasiness over Chinese capacities exist across the region.

Fueling concerns about the possible uses of growing Chinese military and paramilitary capabilities are Beijing's apparent readiness to apply them in territorial disputes coupled with a seeming reticence about subjecting its forces to regulations that can mitigate crises. Beijing has a record of appearing less than enthusiastic over the establishment of dispute management mechanisms relating to a South China Sea Code of Conduct. This despite agreeing to arrangements to implement declarations to put such a code in

---

[1] Minnie Chan and Darren Wee, "Chinese troops join U.S.-Thailand Cobra Gold military exercises," *South China Morning Post,* February 11, 2014, <http://www.scmp.com/news/china/article/1425918/chinese-troopsjoin-asia-us-drills-positive-step-amid-regional-tensions>; Phil Stewart, "China to attend major U.S.-hosted naval exercises, but role limited," *Reuters*, March 22, 2013, <http://www.reuters.com/article/2013/03/22/us-usa-china-drill-idUSBRE92L18A20130322>.

place in 2002, 2011, and 2013.[2] During the Association of Southeast Asian Nations (ASEAN) Ministerial Meetings in 2012, China effectively vetoed intra-ASEAN efforts to develop positions on handling tensions arising from disputes in the South China Sea through Cambodia. These come in addition to recent Chinese efforts use paramilitary forces to challenge Japanese administrative control over the disputed Diaoyu/ Senkaku Islands, assert control over Scarborough Shoal, and disrupt fishing and oil exploration activities by other South China Sea claimants. Chinese forces are apparently taking bolder steps to stop U.S. naval surveillance and other activity in disputed waters that are potentially escalatory.

The rest of the paper seeks to detail how Chinese behaviour sends mixed signals to regional actors about its growing military capabilities. The following section provides an overview of growing Chinese military capabilities and their potential applications. The third section discusses how Chinese behaviour over the past decade created greater apprehensiveness and confusion about growing Chinese capabilities across East Asia, and the re-

---

[2] Carlyle A. Thayer, "New Commitment to a Code of Conduct in the South China Sea?" *National Bureau of Asian Research*, October 9, 2013,
<http://nbr.org/research/activity.aspx?id=360>; Association of Southeast Asian Nations (ASEAN) Secretariat, "Declaration on the Conduct of Parties in the South China Sea," November 4, 2002,
<http://www.asean.org/asean/external-relations/china/item/declarationon-the-conduct-of-parties-in-the-south-china-sea>; ASEAN Secretariat, "Initiatives on South China Sea, Conflict Resolution and Peacekeeping Cooperation, People Engagement, Among Many Issues Noted in Comprehensive ASEAN Joint Communiqué Bali, Indonesia," July 20, 2011,
<http://www.asean.org/communities/asean-political-security-community/item/initiatives-on-south-china-sea-conflictresolution-and-peacekeeping-cooperation-people-engagement-among-many-issues-noted-in-comprehensive-asean-joint-communique-bali-indonesia-20-july-2011>; ASEAN Secretariat, "Terms of Reference of the ASEAN-China Joint Working Group on the Implementation of the Declaration on the Conduct of Parties in the South China Sea,"
<http://www.asean.org/asean/externalrelations/china/item/terms-of-reference-of-the-asean-chia-joint-working-group-on-the-implementation-of-thedeclaration-on-the-conduct-of-parties-in-the-south-china-sea>.

sulting reactions from regional actors. A fourth section highlights regional responses to the uncertainties surrounding China. Section five examines possibilities for overcoming current regional concerns over Chinese intention. The conclusion puts forward some thoughts on the different trajectories for regional politics that may result from current developments in Chinese capabilities and behaviour. For the purposes of this paper, regional actors will include governments in North and Southeast Asia as well as other parties with active interests and participation in the region. This includes Australia, India, New Zealand, Russia, and the United States.

Ⅰ. Growing Chinese Military Capabilities

China's increase in defence spending shows few signs of abating. Since the mid-1990s, Beijing's annual defence expenditure increased from US\$ 33.5 billion in 2000 to US\$ 129.2 billion in 2011 according to the Stockholm International Peace Research Institute (SIPRI).[3] This investment in defence shows clear results in terms of Chinese capabilities. China's navy can now operate away from its shores more effectively than ever before, regularly deploying as far the Gulf of Aden for anti-piracy patrols. China likewise demonstrates significant improvements in conventional airpower just as it apparently attained particular competencies in ballistic missile, anti-satellite, surveillance, and cyber-warfare technologies.[4] These developments are largely unsurprising given China's growing ability to fund key capabilities identified by its military and political leaders, as well as areas where it has longstanding competencies.

---

[3] "China," *The SIPRI Military Expenditure Database,*
<http://milexdata.sipri.org/result.php4>.

[4] U.S. Department of Defense, "Annual Report to Congress: Military and Security Developments Relating to the People's Republic of China 2013,"
<http://www.defense.gov/pubs/2013_china_report_final.pdf>.

Augmenting China's defence capabilities are improvements in its para-military units, particularly those overseeing maritime issues. Last year, China brought several disparate maritime administration and law enforcement arms under the umbrella of a newly formed coast guard under the State Oceanographic Administration.[5] This change addresses operational redundancies and inefficiencies arising from apparent poor coordination among China's maritime agencies in previous years. Apart from administrative improvements, China's maritime agencies are undergoing enhancements in equipment that include more sophisticated, higher displacement ships that can undertake more extensive, complex, and longer range missions.[6] On top of these surface capabilities are the additions of advanced aircraft that support surface missions and can engage in longer range surveillance.

More expansive Chinese capabilities do not have to be inherently destabilising. Improved ability to patrol waters like the South China Sea and navigable rivers that traverse the borders of China and the various Mainland Southeast Asian states can provide greater security to commercial shipping and help maintain the freedom of navigation. The South China Sea is, after all, historically prone to piracy. The last spike in pirate activity there being as recent as the late 1990s and early 2000s.[7] Riverine patrols by Chinese Border Police along the upper reaches of the Mekong River help to prevent at-

---

[5] Christian Le Miere,"Coast Guard Competition in Asia," *Institute of International Strategic Studies*, September 9, 2013,
<http://www.iiss.org/en/militarybalanceblog/blogsections/2013-1ec0/ september-2013-1aeb/coast-guard-china-da6b>; Megha Rajagopalan, "China's civilian fleet plays key role in Asia's disputed seas," *Sydney Morning Herald*, March 6, 2014, < http:// www.smh.com.au/world/chinas-civilian-fleet-plays-key-role-in-asias-disputed-seas-201403 07hvgdv.html#ixzz2z8DTwZCF>.

[6] Le Miere, "Coast Guard Competition in Asia"; Rajagopalan, "China's civilian fleet plays key role in Asia's disputed seas".

[7] David Rosenberg, "The Political Economy of Piracy in the South China Sea," *Naval War College Review*, Vol. 62, No. 3 (2009), pp. 43-58.

tacks on merchant shipping, which in 2011 saw the tragic murder of 13 Chinese civilian sailors.[8]

Chinese force projection capabilities which come with the acquisition and deployment of aircraft carriers and their associated battle groups can help with humanitarian and disaster relief, particularly given the frequency of natural disasters like typhoons and earthquakes in the region. Better satellite surveillance can aid search, rescue, and recovery efforts, as the recent and tragic disappearance of Malaysian Airlines Flight MH370 has shown.[9] Enhancements in Chinese strategic deterrence capabilities in the forms of more advanced long-range ballistic missiles and ballistic missile submarines too can be helpful in preventing major power conflict and limiting the potential for crisis escalation among major powers.[10] Such improvements are perhaps even more prominent than China's more potentially destabilizing and escalatory developments in anti-ship ballistic missile and anti-satellite capabilities.[11]

---

[8] Edward Wong, "China and Neighbors Begin Joint Mekong River Patrols," *New York Times*, December 10, 2011, <http://www.nytimes.com/2011/12/11/world/asia/china-and-neighbors-begin-joint-mekong-riverpatrols.html?_r=0>.

[9] Paul Farrell, "Flight MH370: French and Chinese satellite images show 'potential objects' ," *The Guardian*, March 23, 2014, <http://www.theguardian.com/world/2014/mar/23/mh370-poor-visibility-indian-ocean-searc heffort>.

[10] Thomas J. Christensen, "The Advantages of an Assertive China," *Foreign Affairs*, March/April 2011; Alastair Iain Johnston, "How New and Assertive is China's New Assertiveness," *International* Security, Vol. 37, No. 4 (Spring 2013), pp. 17-20.

[11] Marc Kaufman and Dafna Linzer, "China Criticized for Anti-Satellite Missile Test," *Washington Post*, January 19, 2007, <http://www.washingtonpost.com/wp-dyn/content/article/2007/01/18/AR2007011801029.h tml>; Andrea Shalal, "Analysis points to China's work on new anti-satellite weapon," *Reuters*, March 17, 2014, <http://www.reuters.com/article/2014/03/17/us-chinaspace-report-idUSBREA2G1Q32014 0317>; U.S. Naval Institute, "Report: Chinese Develop Special "Kill Weapon" to Destroy

Yet, it is clear that regional actors in the Asia-Pacific find it difficult to see China's build-up in capabilities in a more benign light. This despite the high and rising levels of economic integration with China, and the common expectation that Chinese military and paramilitary capabilities will naturally improve so long as China's economy and budgets continue to expand. Moreover, less powerful regional actors seem unready to free ride on Chinese largesse in providing public goods such as freedom of navigation in the same way that they traditionally react to American military might in the area. Rather, regional actors appear to be increasingly uncertain of China and Chinese military and paramilitary capabilities even as they seek further cooperation, especially over economic issues. This, of course, begs the question, why?

II . Asia's Uncertainty over China

Uncertainty over advances in Chinese military and paramilitary capabilities among China's neighbours reflects trepidation over Beijing's intentions. This stems from the relative inconsistency of the Chinese government's approach toward its neighbours, magnified by China's economic, political, and sheer physical heft. Even as Beijing pursues more extensive economic integration and diplomatic cooperation with its neighbours, the Chinese government continues to appear resistant to binding mechanisms to manage friction, especially over the many territorial disputes it is party to. Beijing too seems ready to apply its expanding capabilities to press for advantage in these disputes, even with longstanding partners. Expanding Chinese military and paramilitary capacities look to enable Beijing to at times traditional Great Power behaviour, particularly toward weaker neighbours. This worries regional actors in spite of economic gains.

U.S. Aircraft Carriers," March 31, 2009, <http://www.usni.org/news-and-features/chinese-kill-weapon>.

China's willingness to assert maritime claims disputed with other regional actors appear to correspond with the growth in its military and paramilitary capabilities. Since 2008, China has become particularly active in
demonstrating an ability to project force into the East and South China Seas
and beyond.[12] Chinese actions include arrests of Vietnamese fishermen in
disputed waters.[13] Chinese ships allegedly cut the towed sensor array of a
Vietnamese geological survey vessel operating in the South China Sea in
2011 and openly disrupted the operations of the U.S. Navy surveillance vessel, the USNS Impeccable, also in disputed waters.[14] Then there are the
standoffs between China and the Philippines over Scarborough Shoal and
Second Thomas Shoal involving ships from China's Maritime Surveillance
Agency and Fisheries Administration ongoing since 2012, and resulting in
effective Chinese control over the former.[15] 2012 also saw a new Chinese
Oceanographic Administration vessel sail around the southern reaches of

---

[12] Donald K. Emmerson, "China's 'frown diplomacy' in Southeast Asia," *Asia Times*, October 5, 2010, <http:// www.atimes.com/atimes/China/LJ05Ad02.html>.

[13] Jeremy Page, "Hanoi says Chinese shot at boat," *Wall Street Journal*, March 26, 2013,
<http://online.wsj.com/news/articles/SB10001424127887324789504578383820145232386
>; "China Defends Arrest of Vietnamese Fishermen Near Disputed Islands," *Voice of
America*, March 22, 2012,
<http://blogs.voanews.com/breaking-news/2012/03/22/china-defends-arrest-of-vietnamese-
fishermennear-disputed-islands/>; "Vietnam fisherman says beaten during China arrest,"
AsiaOne, April 23, 2012, <http://news.asiaone.com/News/AsiaOne+News/Asia/Story/
A1Story20120423-341608.html>.

[14] "Pentagon says Chinese vessels harassed U.S. ship," *CNN*, March 9, 2009,
<http://edition.cnn.com/2009/ POLI-
TICS/03/09/us.navy.china/index.html?_s=PM:POLITICS>; "Vietnam accuses China in
seas dispute," *BBC Online*, May 20, 2011,
<http://www.bbc.com/news/world-asiapacific-13592508>.

[15] Jason Miks, "China, Philippines in Standoff," *The Diplomat*, April 11, 2012,
<http://blogs.voanews.com/ break-
ing-news/2012/03/22/china-defends-arrest-of-vietnamese-fishermen-near-disputed-islands/
>.

China's claims in the South China Sea to demonstrate Chinese sovereignty, following a port call in Singapore.[16]

The East China Sea too had Chinese demonstrations of capability that unsettled other regional actors. The most visible of these were repeated efforts by Chinese Coast Guard ships and aircraft to challenge Japanese effective control of the Diaoyu/ Senkaku Islands through attempts to enter waters near or overfly the disputed islands in 2012 and 2013.[17] China also established an in late 2013 Air Defence Identification Zone (ADIZ) that covers the disputed territory that goes beyond most other ADIZs in requiring advance reporting by aircraft flying through the zone rather than only if they are heading toward Chinese territory.[18] The ADIZ's coverage means that China may intercept Japanese and other aircraft heading toward the Diaoyu/ Senkaku Islands without prior notification, since Beijing sees them as Chinese territory. Add to these Chinese naval vessels passing close to Japanese waters unannounced, getting into near collisions with U.S. Navy vessels, locking onto Japanese Maritime Self-Defence Force ships with fire control radar, and allegedly shadowing U.S. Navy ships.[19]

---

[16] 徐慧芬，〈中國海巡 31 號船出訪途中將巡航南海維護主權〉，《新浪軍事》，2011 年 6 月 16 日，<http:// mil.news.sina.com.cn/2011-06-16/0804652245.html>。

[17] 蒙克，〈透視中國：中國進逼釣魚島出乎日本意料〉，《BBC 中文網》，2013 年 12 月 18 日，
<http://www.bbc.co.uk/zhongwen/simp/china_watch/2013/12/131217_china_watch_hitoshi_tanaka.shtml>；〈2013 年 1 月 28 日外交部發言人洪磊舉行例行記者會〉，《中華人民共和國外交部》<http:// www.fmprc.gov.cn/mfa_chn/fyrbt_602243/t1008571.shtml>。

[18] Peter A. Dutton, "Caelum Liberam: Air Defense Identification Zones Outside Sovereign Airspace," *The American Journal of International Law*, Vol. 103, No. 4 (October 2009), pp. 691-709; "Announcement of the Aircraft Identification Rules for the East China Sea Air Defense Identification Zone of the PRC," *Ministry of National Defense, People's Republic of China*, November 23, 2013,
<http://eng.mod.gov.cn/Press/2013-11/23/content_4476143.htm>.

[19] Kirk Spitzer, "China Finds a Gap in Japan's Maritime Chokepoints," *Time*, July 18, 2013, <http://nation.time.com/2013/07/18/china-finds-a-gap-in-japans-maritime-chokepoints/>; Barbara Starr, "U.S., Chinese warships come dangerously close," *CNN*, December 13,

Territorial disputes aside, Beijing's eagerness to display newly developed capabilities can potentially appear escalatory and destabilising. Most prominent were the People's Liberation Army's shooting down of a disused satellite as a public test of its new anti-satellite capabilities.[20] Then there is the PLA's much-discussed anti-ship ballistic missile capacity.[21] Behind these demonstrations of capability may be an effort by Beijing to highlight its ability to deter American intervention in conflicts over Taiwan or other disputes, but such behaviour lack the credible assurance necessary for effective deterrence. Consequently, the United States may find these moves provocative and the use of such capabilities highly escalatory in a crisis situation. That such conditions are likely to prove more broadly destabilising puts regional actors on edge.

China's overt display of the more muscular aspects of its newfound military capabilities can discount the contributions China can make toward bringing the Asia-Pacific together for other regional actors. Developing Chinese military capabilities can greatly augment regional efforts at addressing non-traditional security threats, notably over human security. China's navy has played a central role in anti-piracy operations around the Horn of Africa since 2008, while the People's Liberation Army and People's

---

2014, <http:// edition.cnn.com/2013/12/13/politics/us-china-confrontation/>; "4 Chinese naval vessels sail just outside Japanese waters near Okinawa," *Kyodo News International*, March 10, 2014, <https://english.kyodonews.jp/news/2014/03/278425.html>; "China military officials admit radar lock on Japanese ship, says report," *South China Morning Post*, March 18, 2013,
<http://www.scmp.com/news/china/article/1193600/china-military-officials-admit-radar-lo ck-japanese-ship-saysreport>; "Report: Chinese ships confronted Kitty Hawk," Navy Times, January 15, 2008,
<http://www.navytimes.com/article/20080115/NEWS/801150314/Report-Chinese-ships-co nfrontedKitty-Hawk>.

[20] Kaufman and Linzer, "China Criticized for Anti-Satellite Missile Test," Shalal, "Analysis points to China's work on new anti-satellite weapon".

[21] U.S. Naval Institute, "Report: Chinese Develop Special "Kill Weapon" to Destroy U.S. Aircraft Carriers".

Armed Police have substantial experience in peacekeeping operations from East Timor to the Middle East.[22] The emergency air- and sea-lift capacities China deployed to Egypt to evacuate personnel from Libya in addition to China's newly commissioned hospital ship, the Peace Ark, indicate that China's emergent capabilities are readily applicable to HADR missions.[23] Yet, these seem to have a limited effect on soothing public perceptions about China around much of East Asia.

Beijing's apparent flaunting of its advanced capabilities as well has the effect of dampening the assuring effects of Chinese diplomatic and political overtures to actors in East Asia. Beijing's seeming willingness to apply its growing military capabilities to back claims, especially over disputed maritime territory, dulls claims about its willingness to manage differences with neighbours through regional mechanisms. ASEAN members that have maritime disputes with China, in particular, are less sure of Beijing's commitment to moving toward a Code of Conduct that can handle dispute-related

---

[22] Alison A. Kaufman, *China's Participation in Anti-Piracy Operations Off the Horn of Africa: Drivers and Implications* (Arlington, VA: Center for Naval Analyses, 2009); Ben Yunmo Wang, "The Dragon Brings Peace? Why China Became A Major Contributor To United Nations Peacekeeping," *Stimson Spotlight*, July 12, 2013,
<http://www.stimson.org/spotlight/the-dragon-brings-peace-why-china-became-a-major-contributorto-united-nations-peacekeeping-/>.

[23] Gabe Collins and Andrew Erickson, "China dispatches warship to protect Libya evacuation mission: Marks the PRC's first use of frontline military assets to protect an evacuation mission," *China SignPost*, February 24, 2011,
<http://www.chinasignpost.com/2011/02/24/china-dispatches-warship-to-protect-libya-evacuationmissi-
si-
on-marks-the-prcs-first-use-of-frontline-military-assets-to-protect-an-evacuation-mission/≥;
"Chinese hospital ship Peace Ark's on 'friendship' mission to Philippines," *South China Morning Post*, November 21, 2013,
<http://www.scmp.com/news/asia/article/1362240/chinese-hospital-shippeace-arks-friendship-mission-philippines>.

crises when they arise, China's repeated agreements to do so notwithstanding.

Beijing's ADIZ declaration in November 2013 likewise cast doubt on the sincerity of the Chinese Communist Party's Central Committee work plan on enhancing cooperation with China's neighbours announced by President Xi Jinping barely a month before.[24] Even Beijing's use of advanced satellite imaging capabilities in search for the missing Malaysian Airlines aircraft raised concerns in some quarters of Southeast Asia.[25] That East Asian actors are now more circumspect about Chinese diplomatic efforts is ironic given Beijing's past success in winning over neighbours through support for cooperative regional frameworks for much of the early 2000s.[26]

III. Tentative Responses from an Uncertain Region

Mixed signals from Beijing, in part due to its force modernisation efforts, prompt a variety of responses from states active in East Asia. Regional reactions include attempts to involve major and middle powers such as the United States, India, Russia, Japan, Australia, and even the European Union in both traditional and nontraditional security issues that affect regional actors. These include the bolstering of defence ties and alliances. There are as

---

[24] 〈習近平：讓命運共同體意識在週邊國家落實生根〉,《新華網》2013 年 10 月 25 日,<http://news.xinhuanet.com/politics/2013-10/25/c_117878944.htm>。

[25] James Areddy, Richard C. Paddock, and Daniel Stacey, "Jet Search Tests Beijing's Crisis Playbook," *Wall Street Journal*, April 15, 2014, <http://online.wsj.com/news/articles/SB10001424052702303663604579503381216969684?mg=reno64-wsj&url=http%3A%2F%2Fonline.wsj.com%2Farticle%2FSB10001424052702303663604579503381216969684.html>; Kirk Semple and Eric Schmitt, "China's Actions in Hunt for Jet Are Seen as Hurting as Much as Helping," *New York Times*, April 14, 2014, <http://www.nytimes.com/2014/04/15/world/asia/chinas-effortsin-hunt-for-plane-are-seen-as-hurting-more-than-helping.html?_r=0>.

[26] Christensen, "The Advantages of an Assertive China," Emmerson, "China's Frown Diplomacy".

well attempts to build-up indigenous defence capabilities, and even to respond in kind to apparent pressure from Chinese military and paramilitary
forces. These occur concurrently with ongoing engagement of China in a
range of regional cooperative frameworks and economic integration with the
Chinese economy. Broadly termed "hedging", this set of somewhat divergent
policies aim to mollify Beijing while preparing for greater contentiousness,
but can also contribute to greater confusion over intentions and increase the
likelihood of misperception and miscalculation.[27]

One East Asian response to the development of Chinese capabilities is a
continuation of efforts to draw China into cooperative regional frameworks
with the aim of giving Beijing a stake in maintaining the current regional
status quo. Engagement is perhaps most evident in Southeast Asia, seen in
efforts to bring China into processes such as ASEAN Plus Three, ASEAN
Plus One, and talks over a South China Sea Code of Conduct. China and
ASEAN are also party to the China-ASEAN Free Trade Agreement
(CAFTA), formalising the fact that China is now among the largest trading
partners for all the ASEAN members.[28] Cooperative mechanisms are less
common in Northeast Asia, but then there is the nascent China-South Ko-

---

[27] John Hemmings, "China: America Hedges Its Bets," *The National Interest*, December 6,
2013, <http:// nationalinterest.org/commentary/china-america-hedges-its-bets-9510>;
ChengChwee Kuik, Nor Azizan Idris and Abdul Rahim Mohammad Nor, "The China Fac-
tor in the U.S. "Reengagement" With Southeast Asia: Drivers and Limits of Converged
Hedging," *Asian Politics and Policy*, Vol. 4 No. 3 (July 2012), pp. 315-44; Evan S. Me-
dieros, "Strategic Hedging and the Future of Asia-Pacific Stability," *Washington Quarterly*,
Vol. 29, No. 1 (Winter 2005-6), pp. 145-67; Randall L. Schweller, "Bandwagoning for
Profit: Bringing the Revisionist State Back In," *International Security*, Vol. 19, No. 1
(Summer 1994), pp. 72-107.

[28] Lijun Sheng, "China-ASEAN Free Trade Area: Origins, Developments, and Strategic Mo-
tivations," *ISEAS Working Paper: International Politics and Security Issues*, Series No. 1
(2003); ASEAN Secretariat, "ASEAN - China Free Trade Area,"
<http://www.asean.org/asean/external-relations/china/item/ asean-china-free-trade-area>.

rea-Japan Trilateral.[29] Then there are also arrangements like the ASEAN Regional Forum (ARF), Asia-Pacific Economic Cooperation (APEC) forum, and the upcoming Regional Cooperative Economic Partnership (RCEP) that provide platforms for China to work with neighbours across East Asia.

Accompanying regional economic integration efforts are attempts to actively include China in region-wide military and other exchanges that aim to build confidence with Beijing. These include bringing Chinese forces into longstanding military exercises involving other major powers active in the region, such as Cobra Gold in Thailand and RIMPAC.[30] Such interaction comes on top of regular meetings among top military officers and defence officials from around region, such as at the annual Shangri-La Dialogues held in Singapore and joint training sessions for senior military and security officials around the region.[31] There are likewise periodic region-wide search-and-rescue as well as HADR exercises that bring together military, paramilitary, and law enforcement agencies in the attempt to foster greater mutual understanding.[32] An aim of these events is to create greater mutual trust in order to lower the likelihood of miscalculation even as Beijing seems ever more willing and able to bring force to bear on disputes.

---

[29] "Asia's Trilateral Trade Talks," *New York Times*, March 6, 2014, <http://www.nytimes.com/2014/03/07/ opinion/asias-trilateral-trade-talks.html>; Japanese Ministry of Foreign Affairs, "Japan-China-ROK Trilateral Summit," <http://www.mofa.go.jp/region/asia-paci/jck/summit.html>.

[30] Chan and Wee, "Chinese troops join U.S.-Thailand Cobra Gold military exercises," Stewart, "China to attend major U.S.-hosted naval exercises".

[31] Joshua Kurlantzick, "The Shangri-La Dialogue: A Wrap-up," *Council on Foreign Relations: Asia Unbound*, June 3, 2013, <http://blogs.cfr.org/asia/2013/06/03/the-shangri-la-dialogue-a-wrap-up/>.

[32] Scott Cheney-Peters, "Southeast Asian Rivals Work Together in Disaster Relief Exercise," *U.S. Naval Institute News*, June 24, 2013, <http://news.usni.org/2013/06/24/southeast-asian-rivals-work-together-indisaster-relief-exe rcise>.

Bilateral efforts to work alongside China and increase Beijing's stake in the current regional order also exist. Singapore and New Zealand have bilateral free trade agreements with China, while Taiwan has an Economic Cooperation Framework Agreement (ECFA). Japan and Korea too are studying the feasibility of freeing up trade with China. Militarily, there are a number of bilateral exercises involving either various services of the PLA or the Chinese Coast Guard and militaries as well as other forces from Singapore, Malaysia, Thailand, and Australia. [33] They largely centre on counter-terrorism, anti-piracy, and maritime search-and-rescue, which allow militaries and commanders to familiarise themselves with each other's operational approaches in ways that can encourage cooperation and help avoid miscalculation. These come on top of regular bilateral military exchanges and meetings between China and countries from South Korea and Japan through Indonesia, Malaysia, Singapore, Thailand, Vietnam, and even the United States.

Yet, security-related cooperative arrangements in both northeast and southeast Asia concurrently indicate a deep sense of unease about China and its growing capacity to use force. U.S. Deputy Assistant Secretary of State Daniel Russel testified to Congress that the United States will stand by its commitments to oppose Chinese attempts to seize disputed territories in East Asia by force even as it is raising its naval presence in the region.[34] At the

---

[33] Phillip Wen and Mark Kenney, "Tony Abbott's China visit nets closer military relations," *Sydney Morning Herald*, April 13, 2014,
<http://www.smh.com.au/world/tony-abbotts-china-visit-nets-closer-military-relations-201
40412-36jz8.html>; "China active in Southeast Asian joint exercise for 1st time," Asahi *Shimbun*, February 19, 2014,
<http://ajw.asahi.com/article/asia/around_asia/AJ201402190053>; "China, Singapore conduct joint training exercise," *China Daily*, November 19, 2010,
<http://www.chinadaily.com.cn/china/2010-11/19/content_11580752.htm>.

[34] Daniel R. Russell, Assistant Secretary, Bureau of East Asian and Pacific Affairs, "Testimony Before the House Committee on Foreign Affairs Subcommittee on Asia and the Pa-

same time, the United States is publicly reaffirming its alliances with Japan and the Philippines, stating that America's security commitments cover islands under dispute with China.[35] These come on top of long-term rotational deployments of U.S. naval forces to Singapore and Marine forces to Australia, as well as an upgrading of U.S. defence ties between Vietnam and the Philippines. Southeast Asian claimants to the South China Sea are also discussing further mutual coordination between their military and paramilitary forces, which may suggest a common worry about China. Most obvious are recent interactions between the Philippines and Vietnamese militaries on islands disputed with each other and China.[36]

Regional militaries are also upgrading and modernising their own capabilities. The Stockholm International Peace Research Institute (SIPRI) lists Asia as having the fastest rate of defence budget growth in the world, with five of the fifteen top defence spenders coming from Asia in 2013.[37] Outside of North America, Asia also spent the most on its military in 2013. Much of the spending seems aimed at building capacities to assert maritime claims and the freedom of navigation, which seem to parallel Chinese efforts to do the same. These include an investment in submarines, advanced surface ships, and new aircraft by governments in Indonesia, Japan, Malaysia, Sin-

---

cific," Washington, DC: February 5, 2014,
<http://www.state.gov/p/eap/rls/rm/2014/02/221293.htm>.

[35] Manuel Mogato, "U.S. Admiral Assures Philippines of Help in Disputed Sea," *Reuters*, February 13, 2014,

<http://www.reuters.com/article/2014/02/13/us-philippines-usa-southchinasea-idUSBREA1C 0LV20140213>; Russell, Assistant Secretary, "Testimony Before the House Committee on Foreign Affairs Subcommittee on Asia and the Pacific".

[36] Carlyle Thayer, "Is a Philippine-Vietnam Alliance in the Making?" *The Diplomat*, March 28, 2014, <http:// thediplo-mat.com/2014/03/is-a-philippine-vietnam-alliance-in-the-making/>.

[37] "Military Expenditure," *SIPRI Yearbook 2013* (Stockholm, Sweden: Stockholm International Peace Research Institute, 2013), <http://www.sipri.org/yearbook/2013/03>; "Trends in World Military Expenditure, 2012," *SIPRI Fact Sheet*, April 2013, p. 2.

gapore, Taiwan, and Vietnam. Even the cash-strapped Armed Forces of the Philippines is looking to upgrade its air and sea power.[38] Such investments in capability appear to focus on an ability to make assertions of control over regional waters, a mission that China seems keenly interested in developing, highly costly even if they do not enable regional states to project force in any serious manner.

Differentiated regional responses to China's growing capabilities, including "hedging", can complicate regional politics and raise uncertainty. "Hedging" may, in fact, mask a series of security dilemmas in the region. Regional efforts to bolster defence capabilities and security relations with the United States as a sort of insurance policy against fears of potential Chinese aggression alongside American rebalancing to Asia may feed Beijing's worries about containment. Such concerns on the part of China may fuel further capability enhancement and pre-emptive establishment of control over disputed territory that unsettle various regional actors even more. Overtures to China that appear to exclude the United States can prompt more vigorous American efforts to demonstrate interest and commitment to regional security that make Beijing more distrustful of U.S. intentions.

Given that "hedging" is a vague and under-specified concept implies that policies that follow may inadvertently include a range of difficult to discern signals.[39] Rivals may see efforts to simultaneously cooperate while preparing for the possibility of some sort of armed confrontation as indicating duplicitousness. Competitors and disputants may believe cooperation to

---

[38] Richard Jacobson, "Modernizing the Philippine Military," *The Diplomat*, August 22, 2013, <http:// thediplomat.com/2013/08/modernizing-the-philippine-military/>; "Philippines to spend US$524m on military aircraft," *Channel News Asia*, March 22, 2014, <http://www.channelnewsasia.com/news/asiapacific/philippines-to-spend-us/1044382.html>.

[39] Ja Ian Chong, "Lost in transition, or why non-leading powers should concern Beijing and Washington," *East Asia Forum Quarterly*, June 24, 2010, <http://www.eastasiaforum.org/2010/06/24/lost-in-transition-orwhy-non-leading-powers-should-concern-beijing-and-washington/>.

be an attempt to lull them into complacency and erode their positions. Allies can read insufficient commitment into what may look like overly pliant and accommodating policies toward possible rivals. In short, if "hedging" is pairing cooperation with actions that enable more effective use of force in a contingency, it is not a risk- or cost-free strategy. Trying to have your cake and eat it at the same time can have its downsides, much like any other policy. Regional states would do well to recognise that such challenges are possibilities for regional responses to China's growing military and paramilitary capabilities, especially given the high degree of uncertainty already extant in East Asia.

IV. Assuaging Uncertainty and Strategic Mistrust

Given that the development of capabilities by China's military and paramilitary forces are unlikely to abate as the country grows in economic heft, the key question for regional capitals, including Beijing, is how to mitigate the risks associated with this development. Ideally, China can demonstrate commitment to greater strategic transparency and restraint. This can entail more openness about budgets and development programmes for its military as well as paramilitary forces, while spearheading the development of binding dispute and crisis management mechanisms. Being the actor with the largest and fastest growing capabilities in the region, such acts of obvious self-restraint will complement current cooperative arrangements and go a long way in assuring other regional actors. That said, such initiatives put a lot of responsibility on a Chinese government that continues to face significant challenges in domestic economic and administrative reform. Regional actors in East Asia obviously can do more to reduce uncertainty and make conditions more conducive for China to take the lead in the areas mentioned above.

One area regional states can work on is to find common ground on a set of guidelines for dispute and crisis management that they can put in practice.

A set of protocols can reduce the risk of escalation and even limit the use of coercive force over disputes without prejudicing any party's claim. Of course, regional governments with competing maritime claims that do not involve China can go even further by seeking third party through mechanisms like the International Court of Justice or the Law of the Sea Convention. Such moves can help realise the ideas behind Conduct of Parties in the South China Sea, which almost all claimants and interested parties support in principle anyway. The key is in consistently coordinating behaviour and perhaps extending such practices beyond the South China Sea, creating norms that any actor—including China— may find costly to disrupt but advantageous to conform with. There does not even need to be a formalised agreement, which could prove difficult for various governments to complete given their domestic political climates.

A second related issue regional states in East Asia can focus on is look for ways to further encourage compliance with the dispute and crisis management. This means associating support for dispute and crisis management processes with clear incentives, perhaps including publicizing non-compliance by offending governments and withholding assistance in other areas. Disincentives corresponding to noncompliance with agreed common dispute and crisis management practices may be just as important, and can take the form of limiting diplomatic support on other areas of interest to the non-complying government. Such measures should be applied even-handedly to regional states active in the maritime arena and possibly other areas where regional disputes exist to prevent an inadvertent isolation of any one actor. Development in this direction requires regional states to overcome the collective action problems that currently hinder the development of common approaches and best practices which were apparent in previous ASEAN efforts to reach joint positions on behaviour in the South China Sea.

A third element that can help address the uncertainty surrounding China's growing capability to use force is to for regional states to clearly and consistently encourage greater transparency on decision-making processes. Clarifying how the Chinese government develops policies that affects its regional neighbours, its main considerations it includes in such processes, and the challenges it faces can help assure regional actors over Chinese intentions. Uncertainty over Beijing's intentions may be a key reason why Chinese efforts to further develop its military and paramilitary capabilities worry other regional actors, perhaps more than necessary. China's relative security given its physical and economic size as well as its superiority in terms of the capability to use force relative to other actors in the region makes Chinese initiatives in improving transparency particularly assuring to weaker regional actors. In this regard, regional governments in East Asia can impress on leaders in Beijing that enhancing transparency is in China's long-term interests and can pave the way for less contentious relations with its neighbours.

Lastly, regional actors can clarify their unresolved maritime claims and reduce an additional element of uncertainty that may involve growing Chinese military and paramilitary capabilities.[40] One reason for the persistence of competing claims in the South and East China Seas is the vagueness of the various claims, especially in terms of their conformity with relevant international laws and conventions. The nine-dashed line claim over the South China Sea by China and Taiwan is especially unclear under international law, and this is also the assertion that is at the heart of most unease for regional actors. Since Taiwan and China's claims rest on a similar historical and legal basis, clarification by one or both parties can help introduce greater precision to what is under contention and conditions under which these claimants may use force. Claimants and interested parties in maritime as well as land-based territorial disputes in Asia can reduce confusion by making claims as clear as possible under current international legal frameworks, while encouraging others—including China—to do the same.

---

[40] 〈新加坡：中國在南海爭議上態度不明確〉,《德國之聲》, 2011 年 6 月 20 日,
  <http://www.dw.de/新加坡中國在南海爭議上態度不明確/a-15174507-1>。

V. Conclusion

In themselves, China's military and paramilitary capability development are far less problematic than the uncertainty that surrounds them. Much of this has to do with the mixed signals that Beijing seems to be sending to the region about how it intends to apply the new capacities that it is acquiring. Even if such messages are inadvertent and reflect Beijing's inattention more than anything else, regional actors have little way to differentiate such behaviour from action that has less benign intent. Insufficient transparency about policy-making processes in Beijing along with China's disproportionate economic influence and military might relative to other actors in East Asia gives further cause for unease about how Beijing may use its newfound capabilities. Such realities easily lead regional governments to more easily discount the gains and even good faith China built-up from long-term economic interaction and diplomatic cooperation.

As a major power with significant capabilities and unresolved territorial disputes with a number of East Asian actors, China's words and deeds are particularly sensitive for both individual governments and the region as a whole. Regional actors are especially likely to discount Chinese intentions given these real and substantial power asymmetries, as well as Beijing's seeming readiness to use its strengths to push for advantage. That a widespread perception about the eagerness of rising powers to challenge and alter the existing status quo and rules of the game exists further heightens anxieties regional actors have of China's growing military and paramilitary capabilities. Consequently, smooth, stable relations with neighbours will depend a lot on China's ability to assure others through consistent, obvious, and deliberate demonstrations of a commitment to transparency and restraint from the use of coercive force. It is such action that can render Chinese capabilities less worrisome to its neighbours, especially over time.

Will It Be Possible To Deter China Into The 2020s?

Richard D. Fisher, Jr.

(Senior Fellow, International Assessment and Strategy Center)

Ⅰ. Introduction

China's broad and accelerating modernization of its military forces, at first seeking to dominate the East Asian region and then to project power globally, sparks considerable anxiety over whether the United States, either alone or with allied support, can continue to deter China from using force to change what Beijing views as an unfavorable "status quo." This anxiety is compounded by the fact that China is not just an emerging military superpower, but is also gaining economic superpower status which in the views of many helps to legitimize its quest for greater military power. China has used its military and economic power to threaten and constrain Taiwan in the 1990s and 2000s and continues to accumulate forces and capabilities for attacking or invading the island democracy. In the current decade China may decide it can use limited force against Japan and the Philippines to enforce maritime area claims, which would damage Washington's regional military leadership while prompting many states to consider strategic deterrent capabilities. Into the 2020s and beyond it will no longer be a question of deterring Chinese military might in Asia alone, but also those Chinese forces that can be deployed globally to influence conflicts that would affect Western or democratic interests.

At the same time the United States displays internal economic-political turmoil that is detracting from its potential to sustain a military edge sufficient to help deter China. Indeed, the Obama Administration gained some credit by pursing its "Rebalance" toward Asia to help quell allied anxieties about China's military growth. But this is now being undermined by doubts, in fact, pointed statements by U.S. officials that it cannot afford the Re-balance.[1] While the Administration remains opaque about its developing military strategies to counter new Chinese capabilities, there is an early em-phasis on non-nuclear armed missiles and new energy weapons. Traditional capabilities to include aircraft carriers, the new F-35 Fifth-Generation tacti-cal fighter, a new stealthy strategic bomber and nuclear attack submarines receive strong political support, but none are safe from budgetary pressures.

In response to these trends, Taiwan, South Korea and soon Japan have or will be building new missile capabilities. Japan and South Korea maintain significant submarine capabilities while both are pursuing options for 5th generation fighters. Taipei desires both but must surmount budgetary and political hurdles to obtain them. Washington does have options for strength-ening its own and regional deterrent capabilities, but does it have the politi-cal will to employ them? It could help South Korea, Japan and Taiwan to more rapidly obtain new missile and energy weapons to bolster non-nuclear deterrents. But it could also revive its own tactical nuclear forces in Asia in cooperation with its allies but this would run counter to its nuclear disarma-ment preferences.

II. China's Growing Challenge—Briefly

The Chinese Communist Party (CCP) leadership is modernizing and building up its People's Liberation Army (PLA) to prepare for success along two vectors. But before describing these vectors it critical to note they stem from the same base motivation: to sustain the dictatorship of power of the CCP. This requires not just the securing of influence and access to all that is

---

[1] On March 4, 2014 Assistant Secretary of Defense for Acquisition Katrina McFarland told a conference that "Right now, the pivot is being looked at again, because candidly it can't happen," see Zachary Fryer-Biggs, " DoD official: Asia pivot 'can't happen' due to budget pressures," *The Navy Times,* March 4, 2014, <http://www.navytimes.com/article/20140304/NEWS05/303040010/DoD-official-Asia-pivot-can-t-happen-due-to-budget-pressures>.

necessary to sustain the CCP's power position, it also requires that threats to that power be isolated and attacked. So the first vector, that of becoming the dominant economic-strategic power in the Greater Asian region, also re- quires the defeat of democratic Taiwan and the displacement of the U.S.-led strategic network in Asia, plus the isolation of Japan and India. At the same time the CCP is moving gradually along a global vector in which it seek first to secure its immediate economic and then strategic interests, which will likely entail a more vigorous confrontation with the democracies as it pro- motes and helps to consolidate other dictatorships.

To achieve its aims the CCP is not just building its military capabilities, but is also expanding its political warfare capabilities. First, China resists calls for military transparency on its part as well as refusing to engage in any arms control talks that might affect its forces, so that it can better limit what is known about its military capabilities and intentions. Since 2003 it has put into practice new strategies called "The Three Warfares."[2] These include legal warfare, such as reinterpreting of the United Nations Law of the Sea Treaty to expand territorial rights within Economic Exclusion Zones and psychological warfare, the use of paramilitary and military forces to intimi- date to intimidate Japan and the Philippines from defending maritime claims. Media warfare is seen in the rapid expansion of China's state-controlled me- dia overseas to try to dominate media markets. China is also using a form of

---

[2] See Dean Cheng, "Winning Without Fighting: Chinese Legal Warfare," *Heritage Founda- tion Backgrounder No. 2692,* May 21, 2012, <http://www.heritage.org/research/reports/2012/05/winning-without-fighting-chinese-legal- warfare>; and more recently, Professor Stepfan Halper, *China: The Three Warfares* (Cam- bridge: University of Cambridge, 2013); For Andy Marshall, Director, Office of Net As- sessment, Office of the Secretary of Defense, Washington, D.C., May 2013, reviewed by Bill Gertz, "Warfare Three Ways, China waging 'Three Warfares' against United States in Asia, Pentagon says," *The Washington Free Beacon,* March 26, 2014, <http://freebeacon.com/national-security/warfare-three-ways/>.

"active measures" targeting political elites in the United States and Taiwan to enlist their support for decisions that contribute to Chinese goals.[3]

CCP success over the long term, of course, is not pre-ordained. It will require not just the promotion of sustained economic growth to encourage a technically innovative and loyal plurality of Chinese, but also continued submission to the world's most sophisticated surveillance-police state security apparatus. Again, CCP success is not assured, both in sustaining economic growth sufficient for quelling political challenge or in sustaining the loyalty of its PLA-Police security system. So the CCP requires that others make consistent errors, such as Taiwan failing to sufficiently strengthen and guard its democracy. It also requires that the other democracies, especially the United States, fail to sustain their global power and vigilance against CCP-inspired threats. Furthermore, it will not be enough to meet the CCP challenge in the Asian region alone, but it will be necessary to confront it outside of Asia as well.

China is already well on its way to assembling coalitions in multiple regions: Shanghai Cooperation Organization in Central Asia and the Forum on China-Africa Cooperation are two it helps to lead, the West-excluding Community of Latin American and Caribbean States (CELAC) is one it supports.[4] Beijing has created examples of nuclear power projection in North Korea, Pakistan and Iran, providing economic props for narrowly based authoritarian-to-dictatorial regimes which it has also enabled to ac-

---

[3] Mark Stokes and Russell Hsiao, "The People's Liberation Army General Political Department, Political Warfare with Chinese Characteristics," *Project2049 Web Page,* October 14, 2013,
<http://www.project2049.net/documents/PLA_General_Political_Department_Liaison_Stokes_Hsiao.pdf>.

[4] In February 2012 a diplomatic source told the author that then CCP leader Hu Jintao called and congratulated the Latin state leaders who attended the first CELAC summit in July 2011.

quire nuclear missiles.[5] It can be expected that China will seek to preserve loyal regimes in these countries. While helping to neutralize India and Japan, this investment provides a stunning example of power projection to aspiring regional powers like Saudi Arabia, Turkey, Brazil, Egypt and Argentina, which have regional power aspirations that could be satisfied by Chinese high-tech weapons.

1. Strategic Heights

To propel its quest for dominant strategic power the CCP is building PLA strength in two key capabilities: nuclear/long-range strike and space control. Though China has never revealed its nuclear arsenal, the longstanding assessment of 300-400 nuclear weapons[6] has been undermined by recent analysis pointing to China's construction of 5,000km of tunnels to conceal its nuclear missiles and weapons.[7] Furthermore, a recent estimate offered by a former Chief of Staff of the Russian Strategic Missile Forces Major General Victor Esin (ret.) puts this number as high as 900 deployed warheads.[8] Esin also estimates China may have 650 tactical nuclear weapons.[9] The soon-to-be deployed multiple-warhead mobile DF-41 ICBM will al-

---

[5] China's record of nuclear and missile technology proliferation is explored by the author in *China's Military Modernization, Building for Regional and Global Reach* (Stanford: Stanford University Press, 2010), pp. 47-58.

[6] Hans M. Kristensen, "STRATCOM Commander Rejects High Estimates for Chinese Nuclear Arsenal," *FAS Strategic Security Blog,* August 22, 2012, <http://blogs.fas.org/security/2012/08/china-nukes/>.

[7] William Wan, "Digging Up China's Nuclear Secrets," *The Washington Post,* November 30, 2011, p. 1.

[8] Colonel General (ret.) Victor Yesin, "Third After the United States and Russia, On China's Nuclear Potential Without Underestimation or Exaggeration," originally published in *Voenno-promyshlenyi Kur'er (VPK),* No. 17 (May 2, 2012), translated by Anna Tsipokina for The Potomac Foundation, with assistance from Dr. Hung Nyugen, p. 2.

[9] Victor Yesin, "Third After the United States and Russia, On China's Nuclear Potential Without Underestimation or Exaggeration," p. 507.

low for a rapid increase in strategic warheads. From underground bases in the Delingha region, four brigades of DF-41s (48 missiles x 10
warheads)[10] could take out the U.S. *Minuteman* ICBM force (to be
reduced to 400 missiles) in a potential first strike. The PLA will soon
deploy a new 3,500-4,000km range DF-26 intermediate range ballistic
missile,[11] in addition to its 100 or so DF-21 family medium range
ballistic missiles, 1,100 or so short range ballistic missiles and 500 or
so land attack cruise missiles.

China began testing its anti-satellite (ASAT) warfare weapons with initial ground laser tests in 2005, laser tests against U.S. satellites in 2006,
tested its SC-19 ASAT missile in January 2007 and in May 2013 tested a
Medium Earth Orbit capable DN-2 ASAT.[12] The dual-use *Shenzhou* and
Tiangong manned space craft will be joined by a dual use space station and
space planes[13] and space-laser battle platforms[14] in the 2020s. China will
also likely give military capabilities to early unmanned an then manned installations on the Moon. By the early 2020s the PLA will have a full 36+

---

[10] A Second Artillery ICBM brigade is estimated to have six launchers plus one reload missile, for a total of 12 missiles. In November 2013 an Asian military source told the author
that it is normal PLA practice for mobile ICBM brigades to have one reload missile per
launcher.

[11] Bill Gertz, "China Fields New Intermediate-Range Nuclear Missile," *The Washington Free
Beacon,* March 3, 2014,
<http://freebeacon.com/china-fields-new-intermediate-range-nuclear-missile/>.

[12] Bill Gertz, "China Conducts Test of New Anti-Satellite Missile," *The Washington Free
Beacon,* May 14, 2013,
<http://freebeacon.com/china-conducts-test-of-new-anti-satellite-missile/>.

[13] For more background on China's space plane ambitions, see author, "China's Space Plane
Program," *International Assessment and Strategy Center Web Page,* July 27, 2011,
<http://www.strategycenter.net/research/pubID.253/pub_detail.asp>.

[14] For an expansive discussion by Chinese experts of space laser weapons for China, see, Gao
Min-hui, Zheng Yu-quan and Wang Zhi-hong (all of the Changchun Institute of Optics, Fi-
ne Mechanics and Physics), "Development of space-bases laser weapon systems," *Chinese
Optics,* December 2013, pp. 810-817.

Compass navigation satellite system plus scores of optical, radar and electronic intelligence satellites. All of this will contribute to a greater capacity for "space control" into the 2020s.

2. Increasing Anti-Access Layers

By 2003 or so it was becoming clear that the PLA was in the midst of a buildup of a space-radar-communications-computers based C4ISR (command, control, communications, computers, intelligence, surveillance and reconnaissance) system that would cue and control a force of ballistic and cruise strike missiles, 4th and then 5th generation combat aircraft, plus modern submarine, surface ship and mine naval forces.[15] Leveraging its geographic proximity, China has formed a "reconnaissance strike complex" to deny access to U.S. and allied forces within the "First Island Chain."[16] This initial strike complex includes DF-21D anti-ship ballistic missiles, YJ-12 and YJ-18 supersonic anti-ship missiles, H-6K, J-16 and JH-7 strike fighters, new unmanned combat aerial vehicles (UCAVs) plus new *Song* and *Yuan* class conventional submarines. A more survivable and larger nuclear force, plus greater capabilities for space control, will add heft to the PLA's expanding "anti-access" capabilities in the "Second Island Chain" out to Guam.

---

[15] These trends were noted by the author in 2005 testimony before the U.S. Congress, "CHINA'S MILITARY POWER: AN ASSESSMENT FROM OPEN SOURCES," Testimony of Richard D. Fisher, Jr., International Assessment and Strategy Center, Before the Armed Services Committee of the U.S. House of Representatives, July 27, 2005, <http://www.strategycenter.net/docLib/20050731_TestRDFHASC072705.pdf> ; these trends were described in fuller detail in Roger Cliff, Mark Burles, Michael S. Chase, Derek Eaton, Kevin L. Pollpeter, *Entering The Dragon's Lair, Chinese Anti Access Strategies and Their Implications for the United States* (Washington, DC: RAND Corporation, 2007), <http://www.rand.org/pubs/monographs/2007/RAND_MG524.pdf>.

[16] For an excellent review of this evolving reconnaissance strike complex, see Ian Easton, "China's Evolving Reconnaissance Strike Capabilities, Implications for the U.S.-Japan Alliance," *Project2049 Institute Web Page,* February 2014, <http://www.project2049.net/documents/Chinas_Evolving_Reconnaissance_Strike_Capabilities_Easton.pdf>.

This emerging new layer of systems will include the DF-26 IRBM modified for ASBM missions, expected long-range stealth bombers and nuclear attack submarine outfitted to carry LACMs or new MRBMs. Then missiles like the DF-26 or larger missiles may also be equipped with maneuvering Hypersonic Glide Vehicle warheads, enabling a PLA version of non-nuclear "Prompt Global Strike."[17] The PLA will also deploy 400km and 500km and longer-range SAMs. Early aircraft carrier battle groups with J-15/J-15S based air wings, along with new air defense escorts like the Type 052D and future Type 055 destroyers, will extend PLA reach into the Second Island Chain.

3. Focus on Taiwan, Maritime Territories

Recent Taiwanese Ministry of National Defense (MND) assessments that the PLA may be able to invade by 2020,[18] and could prepare within four months,[19] should be regarded as a serious warning. The PLA ballistic missile threat to Taiwan is shifting to low-trajectory and maneuverable systems like the DF-12 and DF-16 ballistic missiles and CJ-10 cruise missile, to better defeat Taiwan's missile defenses. PLA adoption of multiple precision guided munition (PGM) families heightens the effectiveness of PLAAF and PLANAF multi-role fighters. Early projections that the PLA Navy will build up to 6 Landing Platform Dock (LPD) and 6 Landing Helicopter Dock (LHD) large amphibious assault ships[20] must also include the potential that scores

---

[17] Richard Fisher, "China confirms hypersonic vehicle test," *Jane's Defence Weekly,* January 15, 2014.

[18] Rich Chang and Jason Pan, "Chinese could be ready to invade by 2020: MND," *The Taipei Times,* March 14, 2014,
<http://www.taipeitimes.com/News/front/archives/2014/03/06/2003584974>.

[19] Rich Chang, "China can invade in months: MND," *The Taipei Times,* April 4, 2014,
<http://www.taipeitimes.com/News/taiwan/archives/2014/04/04/2003587265>.

[20] Prasun K. Sengupta, "Spotlight on China's LPDs, LHDs and Carrier," *Tempur,* July 2008, pp. 91-93.

of large Roll-On-Roll-Off (RoRo) ferries, rail ferries,[21] Pure Car/Truck Carrier (PCTC) ships can transport 5-10 divisions or brigades—dependent upon the capture of Taiwanese ports. Early acquisition of large *Zubr* hovercraft and new Il-76 based refueling aircraft, both from the Ukraine, could help near-term attempts to capture the Senkaku/ Daiyoutia Islands from Japan. Capture of the nearby Sakashima Island Group could position the PLA for multi-axis attacks against Taiwan.

China is also imposing control over most of the South China Sea to create a "bastion" to protect early nuclear missile submarine (SSBN) patrols and to protect access to the North and West for its carrier battle groups based on Hainan Island. The Woody Island airbase, maritime patrol aircraft and underwater sound sensors[22] will form an initial barrier. In the near term China may be tempted to push Chinese and Vietnamese off of their Spratly Islands before they gather new military strength and to undermine U.S. coalition building. In the longer term China's naval and amphibious naval modernization may make possible punitive PLA raids against main islands of the Philippines.[23]

4. Building for Power Projection

A key motivation of the 2004 "New Historic Mission" for the PLA was that the CCP was going to derive an increasing amount of its legitimacy, or a continuation of its dictatorship, on the premise that the Party was best able to secure China's growing international economic-strategic interests. Indeed, as the PLA strengthens its reach into the Second Island, it is also accumulating

---

[21] Nick Brown and Christopher Foss, "Rising STOM: China expands amphibious capabilities," *IHS International Defense Review,* October 13, 2013.

[22] Lyle Goldstein and Shannon Knight, "Wired for Sound in the 'Near Seas,'" *U.S. Naval Institute Proceedings,* April, 2014, pp. 56-61.

[23] For an account of growing Philippine anxieties, see Richard Fisher, "Tensions Between China and the Philippines Grows," *Aviation Week and Space Technology, Defense Technology Edition,* September 3, 2012, p. 22.

the means for more distant power projection. Nuclear powered aircraft carriers, larger 10,000 ton "Type 055" cruisers,[24] 40,000 ton LHDs are likely to be built. These will be joined by 100 or more 65-ton capacity Xian Aircraft Corporation Y-20 strategic transports, and perhaps larger "C-5" size transports into the 2020s.[25] PLA ground forces are now building brigade size units around all-wheeled medium-weight tank, artillery, air defense and support vehicles, which can be carried by the Y-20. The PLA is also developing heavy "quad-rotor" vertical lift transports to enable rapid heavy assault.[26] Beijing will have the option to fan conflicts, for example, by selling advanced weapons to Argentina[27] to "reunify" the Falklands Islands.[28] Should the Argentines win, Beijing would then be rewarded with greater political, economic and military access in Latin America, displacing American power.

---

[24] For an early assessment of the Type 055 cruiser, see, Richard Fisher, James Hardy and Ridzwan Rahmat, "China's new Yuan class sub seen preparing for sea trials," *IHS Jane's Defence Weekly, online edition,* April 9, 2014, <http://www.janes.com/article/36577/china-s-new-yuan-class-sub-seen-preparing-for-sea-tr ials>.

[25] A Chinese aviation official acknowledged the larger transport aircraft ambition to the author at the IDEX arms exhibit in Abu Dhabi in February 2013.

[26] Richard Fisher, "China's Advanced Helicopter Concepts," *International Assessment and Strategy Center Web Page,* September 7, 2013, <http://www.strategycenter.net/research/pubID.317/pub_detail.asp>.

[27] At the June 2013 Paris Airshow, Argentine aircraft officials acknowledged for the first time their discussions with China to co-produce its Chengdu Aircraft Corporation FC-1 fighter, armed with the near-hypersonic speed CM400AKG attack missile. See, Richard Fisher, "Argentine officials confirm joint-production talks over China's FC-1 fighter," *Jane's Defence Weekly,* June 24, 2013.

[28] In early 2013 an Argentine source told the author, that since the 1982 Falklands War, Chinese officials have told their Argentine counterparts that they view Argentina's quest for the Falklands through the same lens as China's pursuit of Taiwan.

III. The American Rebalance or "Pivot"

Increasing Asian regional anxiety caused by China's actions and its galloping military buildup, plus the opportunity afforded by ongoing U.S. withdrawal from Iraq and Afghanistan, together p prompted the Obama Administration's 2010-2011strategy to "rebalance" toward Asia. In her October 2011 *Foreign Policy* magazine article, Secretary of State Hillary Clinton said the U.S. had reached a "pivot point" that required renewed U.S. emphasis on Asia, and while taking care to reject that China is a "threat." She also emphasized that U.S. military "treaty alliances with Japan, South Korea, Australia, the Philippines, and Thailand are the fulcrum for our strategic turn to the Asia-Pacific."[29] This Rebalance included military posturing and an emphasis on trying to engage China, with great care taken to portray U.S. moves as non-threatening to China. Washington's military initiatives have included:

1. U.S. Forces

In June 2012 the U.S. announced it would shift up to 60 percent of U.S. naval forces to the Pacific.[30] New systems like the P-8 *Poseidon* anti-submarine jet and the F-35 stealthy multirole fighter would first be deployed in the Pacific theater. In 2014 the U.S. increased to four the number of nuclear attack submarines based in Guam.[31]

---

[29] Hillary Clinton, "America's Pacific Century," *Foreign Policy Magazine,* November, 2011, <http://www.foreignpolicy.com/articles/2011/10/11/americas_pacific_century>.

[30] Jane Perlez, "Panetta Outlines New Weaponry for Pacific," *The New York Times,* June 1, 2012, <http://www.nytimes.com/2012/06/02/world/asia/leon-panetta-outlines-new-weaponry-for-pacific.html?_r=0>.

[31] "Navy adds 4th submarine to Guam-based fleet," *Associated Press,* February 11, 2014, <http://news.yahoo.com/navy-adds-4th-submarine-guam-based-fleet-215343914.html>.

2. Australia-Singapore

During President Obama's mid-November 2011 visit to Australia it was announced that the U.S. would station/rotate up to 2,500 Marines in Darwin by 2016 and then that the U.S. may rotate up to four Littoral Combat Ships in Singapore.[32]

3. Japan

In late August 2012 it was revealed that the Administration was planning to move a second long-range X-Band radar to one of Japan's southern islands and that it considered placing a third X-Band radar in the Philippines.[33] In 2013 the U.S. deployed 24 MV-22 *Osprey* tilt-rotor assault-transport aircraft to Okinawa and in April 2014 it was announced that two additional missile destroyers will be based in Japan by 2017.[34]

4. Philippines

A dialogue with the Philippines that started in early 2012 is expected to culminate in early 2014 in a new 20-year agreement to allow "rotational" stationing of U.S. forces in Philippine bases,[35] reversing their controversial exit in 1991.

---

[32] Dan Lothian and Lesa Jansen, "Obama pledges U.S. military power in Pacific," *CNN.Com Web Page,* November 16, 2011,
<http://www.cnn.com/2011/11/16/world/asia/australia-obama-trip/index.html?hpt=hp_t2>.

[33] Adam Entous and Julian E. Barnes, "U.S. Plans New Asian Missile Defenses," *The Wall Street Journal,* August 22, 2012.

[34] Mizuho Aoki, "Two Aegis ships bound for Japan," *The Japan Times,* April 6, 2014,
<http://www.japantimes.co.jp/news/2014/04/06/national/two-aegis-ships-bound-for-japan/#.U0rv9vldV0o>.

[35] Barbara Mae Dacanay, "Philippines sets time limit for US bases," *Gulf News,* March 24, 2014,
<http://gulfnews.com/news/world/philippines/philippines-sets-a-time-limit-for-us-bases-1.1308104>.

5. Taiwan

Taipei started operation of a U.S.-supplied long-range radar in 2013 and started taking delivery of P-3 *Orion* anti-submarine patrol aircraft and AH-64 *Apache* attack helicopters, but the Obama Administration has declined to sell Taipei new F-16 fighters.

6. Developing Air Sea Battle

Despite its preoccupation with the War on Terror for the last decade, the Department of Defense's interest in a "pivot" to Asia extends to early in the last decade when it started becoming clear that China was building an "asymmetric" force of space, submarine and innovative anti-ship ballistic missile (ASBM) weapons to execute "anti-access" and "area denial" (A2/AD) strategies against U.S. forces. Early open U.S. recognition of PLA interest in terminally-guided medium range ballistic missile emerged in 1996[36] that would be acknowledged as an anti-ship threat in the 2005 Pentagon China Military Report.[37]

A focus of extensive inter-service and Department of Defense development and review, by early November 2011 the Pentagon announced the formation of its new Air-Sea Battle (ASB) Office[38] that the Obama Administration has gone to some length to deny as "Anti-China,"[39] but instead focused on "generic" anti-access challenges in other regions. Rather than advocate new capabilities, the Air-Sea Battle Office is charged with proposing

---

[36] Author interviews at the November 1996 Zhuhai Airshow, see author's "China's Missile Threat," *Wall Street Journal,* December 30, 1996, p. A12.

[37] *2005 Department of Defense China Military Report*, pp. 4, 33.

[38] Bill Gertz, "Battle Concept Signals Cold War Posture on China," *The Washington Times,* November 10, 2011, p. 13.

[39] Bill Gertz, "Air Sea Battle Fight," *The Washington Times,* October 12, 2011.

ways to increase "jointness" between U.S. military services.[40] The principle strategy for employment of U.S. forces to counter longer-range "anti access" and shorter-range "area denial" threats was outlined in the Department of Defense January 2012 Joint Operational Access Concept (JOAC).[41] Again, not naming China, but listing the kinds of threats its poses, the JOAC outlines goals using the "synergy" gaining superiority in "multiple domains," like air, maritime, space and cyber, to assure access. The JOAC states it may be necessary to attack the enemy's territory, while acknowledging this raises the risk of escalation, and that it may be necessary to attack enemy space and cyber assets.[42] In a May 2013 document the Air Sea Battle Office further refined the challenges pointing out that adversaries would attack by surprise, attacks will be launched against U.S. forces and allies supporting the U.S., and that all domains would be attacked,[43] and offered a "Central Idea:"

> **The ASB Concept's solution to the A2/AD challenge in the global commons is to develop networked integrated forces capable of attack-in-depth to disrupt, destroy and defeat adversary forces (NIA/D3). ASB's vision of networked, integrated, and attack-in-depth (NIA) operations requires the application of cross-domain operations across the interdependent warfighting domains (air, maritime, land, space, and cyberspace), to disrupt, destroy, and defeat (D3)**

---

[40] The leaders of the US Air Force and US Navy explored their ideas about improving jointness in, General Norton A. Schwartz, USAF and Admiral Jonathan W. Greenert, USN, "Air-Sea Battle -- Promoting Stability in an Era of Uncertainty," *The-American-Interest.com*, February 20, 2012, <http://www.the-americaninterest.com/article.cfm?piece=1212 >.

[41] Department of Defense, *Joint Operational Access Concept (JOAC), Version 1.0,* January 17, 2012, *Department of Defense*, <http://www.defense.gov/pubs/pdfs/joac_jan%202012_signed.pdf>.

[42] Department of Defense, *Joint Operational Access Concept (JOAC), Version 1.0*, pp. 24, 26.

[43] Air Sea Battle Office (Department of Defense), "Air-Sea Battle, Service Collaboration to Address Anti-Access & Area Denial Challenges," May 2013, pp. 3, 4.

A2/AD capabilities and provide maximum operational advantage to friendly joint and coalition forces.[44]

A spirited debate has ensued over the Air Sea Battle as it would apply to China. On the one hand the National Defense University's T.X. Hammes advocated a strategy of "Offshore Control" that stresses dominating the First Island Chain using superior U.S. naval forces and access to friendly bases to terminate conflicts quickly, but that China's nuclear forces remove the option of attacking Chinese territory or to threaten the existence of the CCP regime.[45] It is also said that this strategy is also more affordable because it does not rely on new expensive long range strike systems. Others defend the Department of Defense ASB concept as they also defend the necessity to strike Chinese military capabilities in its territory if deemed necessary.[46]

There is also debate over the new military capabilities needed to better fulfill these new strategies. The U.S. Air Force has placed a very high priority on the development of a new stealthy but affordable subsonic manned bomber, perhaps acquiring up to 80 to 100.[47] A recent RAND studies have highlighted the bomber's greater ability to sustain a conflict,[48] and to signal

---

[44] Air Sea Battle Office (Department of Defense), "Air-Sea Battle, Service Collaboration to Address Anti-Access & Area Denial Challenges," p. 4.

[45] Col. T.X. Hammes (ret.), "Offshore Control Is The Answer," *U.S. Naval Institute Proceedings,* December 2012,
<http://www.usni.org/magazines/proceedings/2012-12/offshore-control-answer>.

[46] Elbridge Colby, "Don't Sweat Air Sea Battle," *The National Interest,* July 31, 2013,
<http://nationalinterest.org/commentary/dont-sweat-airsea-battle-8804?page=2>.

[47] Aaron Mehta, "USAF Defends Need for New Long Range Bomber," *Defense News,* February 20, 2014,
<http://www.defensenews.com/article/20140220/DEFREG02/302200043/USAF-Defends-Need-New-Long-Range-Bomber>.

[48] Thomas Hamilton, *Expendable Missiles vs. Reusable Platform Costs and Historical Data,* Rand Corporation, Project Air Force, 2012,
<http://www.rand.org/content/dam/rand/pubs/technical_reports/2012/RAND_TR1230.pdf>
.

and control escalation over that of new missiles.[49] The U.S. Navy is more interested in equipping new attack submarines with up to 40 cruise missiles or modified Landing Platform Dock ships with 200 or more defensive an offensive missiles. Doing so would seek a more capable/affordable distribution of "fire" capabilities between stealthy and unstealthy "platforms" and new and more capable "payloads."[50] But the U.S. services also seek to preserve existing expensive programs like the Air Force's ambition to purchase over 1,700 stealthy Lockheed-Martin F-35 fighters and the Navy seeks to preserve its major capitol ships of 11 aircraft carrier battle groups and 48 nuclear attack submarines.

IV. How Washington Is Increasing Risk Undermining the Pivot

While still a work-in-progress and at least until 2014, a laudable attempt to assure its Asian allies and friends that it was working to meet the future Chinese challenge, the Obama Administration's rebalance to Asia proceeded in parallel with strategic miscalculations, that when combined with funding limitations, have served to undermine its credibility. By early 2014 mid-level U.S. bureaucrats were sounding the alarm: the Pivot was unaffordable and the U.S. was in danger of losing military-technical competitions with China.

---

[49] Forest E. Morgan, *Crisis Stability and Long Range Strike, A Comparative Analysis of Fighters, Bombers and Missiles,* RAND Corporation, Project Air Force, 2013, <http://www.rand.org/content/dam/rand/pubs/monographs/MG1200/MG1258/RAND_MG 1258.pdf>.

[50] Hon. Robert O. Work, Undersecretary of the Navy, "AirSea Battle: Power Projection in the Mature Guided Munitions Era," Power Point Presentation for the AIE Counter A2/AD Conference, October 26, 2010; a more recent expansion of this theme comes from the Chief of Naval Operations, Admiral Jonathan W. Greenert, U.S. Navy, "Payloads Over Platforms: Charting A New Course," *United States Naval Institute Proceedings,* July 2012, <http://www.usni.org/magazines/proceedings/2012-07/payloads-over-platforms-charting-n ew-course>.

1. Misjudging adversaries

In one sense, the Pivot was an effort to recover from a serious mis-judgment of China and its intentions. The original impetus for the Rebalance of 2010-2011 was the Obama Administration's failed attempt to seek a "partnership" with China. Starting February 2009 President Obama almost immediately reached out to Beijing to try to form what Secretary of State Hillary Clinton called a "comprehensive partnership." This would have entailed downplay of human rights, continued U.S. restraint in its support for Taiwan, while seeking to elevate China in global leadership circles in hopes of enlisting its help in meeting an global and regional concerns like arms control, climate change and preventing North Korea's nuclear ambitions.[51] But by early 2010 the hope for such a partnership had practically vanished. China had passed the point of "joining" the West and instead was intent on advancing its own interests, made clear in its rejection of cooperation on climate change, arms control, plus its defense of North Korea after it sank a South Korean corvette that March 2010. China's rejection of mediation of conflicting claims in the South China Sea at the July 2010 summit of the Association of Southeast Asian Nations (ASEAN) has been followed by more aggressive building of means to "control" this region and "pushing" against the Philippines.

In addition, the Obama Administration has misjudged the determination of Russia to regain its "lost" regions by force of arms. Russian aggression in Europe could quickly deflate the recent U.S. "Pivot" to Asia and increase the threat of Russian and Chinese military coordination and cooperation. A Rus-

---

[51] Glenn Kessler, "Clinton Criticized for Not Trying to Force China's Hand," *The Washington Post,* February 21, 2009,
<http://www.washingtonpost.com/wp-dyn/content/article/2009/02/20/AR2009022000967.h tml> ; David Shambaugh, "Early Prospects of the Obama Administration's Strategic Agenda With China," *Foreign Policy Research Institute Web Page,* April 2009,
<http://www.fpri.org/enotes/200904.shambaugh.obamastrategicagendachina.html>.

sian invasion of the Ukraine and/or the Baltic states could escalate into conventional or even nuclear conflict with NATO, which would involve U.S. forces. A large conflict would quickly re-order U.S. military priorities away from Asia. To both prevent China from exploiting its Western strategic-military preoccupation, and to gain its strategic cooperation, Russia may also accelerate military "coordination" or even cooperation with China, in addition to ramping up arms sales. According to a Chinese account, in 1969 the U.S. threatened nuclear retaliation if the Soviet Union attacked China with nuclear weapons.[52] Would Russia today "tilt" its nuclear weapons against the U.S. in the event China attacked Taiwan? Russian and Chinese military forces have conducted five land and three naval exercises since 2005. Early Russian sales to China may include the 4+ generation Sukhoi Su-35,[53] the 5th generation S-400 SAM,[54] and modern conventional submarine technology. Later sales may include strategic defense and space/Moon exploration technology. Like China, Russia has been developing plans to "colonize" the Moon by mid-century.[55]

---

[52] As relayed in Liu Chenshan, "1969 nian Zhebao Dao chongtu: Sulian yu Zhongguo waike shoushu shi he

daji" (The 1969 Zhenbao Island Conflict: The Soviet Union's Desire to Conduct a Surgical

Nuclear Strike Against China), Wenshi Cankao (Historical Reference), April 30, 2010, cited in Michael S. Gerson, *The Sino-Soviet Border Conflict, Deterrence, Escalation and the Threat of Nuclear War in 1969,* Center for Naval Analysis, November 2010, p. 37, study funded by Defense Threat Reduction Agency, Advanced Systems and Concepts Office.

[53] "The contract between Russia and China for delivery of Su-35 fighters will be signed this year," *Russia Aviation Web Page,* March 19, 2014, <http://www.ruaviation.com/news/2014/3/19/2231/>.

[54] Ivan Safronov, "China is the first customer for the S-400," *Kommersant,* March 28, 2014, <http://saidpvo.livejournal.com/278803.html>.

[55] Russian plans and space craft designs to return to the Moon and go to Mars were revealed at the 2005 and 2007 Moscow Air Shows. Also see, Neil McAllister, "Russian Deputy PM: 'We are coming to the Moon forever,'" *The Register,* April 12, 2014, <http://www.theregister.co.uk/2014/04/12/russia_permanent_moon_base/>.

Defense planning guidance issued by President Obama in January 2012 ended the previous "two-war" defense budget planning guidance, with planning guidance reduced to fight one major conflict and a smaller "holding' conflict.[56] It would appear that Russia and China are both quite sensitive to how reductions in U.S. military strength give them both greater political-military space for maneuver. When adding the uncertainties of a near term nuclear missile-armed Iran and North Korea, there is a growing danger that Russia and China may take advantage of U.S. preoccupation with a major crisis.

2. Nuclear miscalculation

President Obama has made the pursuit of nuclear weapons reductions and the limitation of nuclear proliferation a centerpiece of his foreign policy. Not even two months into his Administration, in Prague on 5 April 2009, Obama declared "America's commitment to seek the peace and security of a world without nuclear weapons."[57] In an April 2010 *Nuclear Posture Review* (NPR) the Administration reduced the role nuclear weapons with new policies: 1) the U.S. "will not threaten" or "use" weapons against non-nuclear states in compliance with the Nuclear Non-Proliferation Treaty; 2) the U.S. would only use nuclear weapons in "extreme circumstances" to defend the "vital interests" of the U.S. "or its allies and partners," ; and 3) strengthen conventional capabilities "and reduce the role of nuclear weapons in deterring non-nuclear attack…to make deterrence of nuclear attack on the United States or our allies

---

[56] United States Department of Defense, "Sustaining U.S. Global Leadership: Priorities For 21st Century Defense," January 2012, p. 4, *Department of Defense*, <http://www.defense.gov/news/Defense_Strategic_Guidance.pdf>.

[57] The White House, "Office of the Press Secretary, Remarks by President Barack Obama, Hradcany Square, Prague, Czecholovakia," April 5, 2009, *The White House*, <http://www.whitehouse.gov/the_press_office/Remarks-By-President-Barack-Obama-In-Prague-As-Delivered>.

and partners the sole purpose of U.S. nuclear weapons."[58] President Obama advanced his goals by reaching the New Strategic Arms Reduction Treaty (New START) on 8 April 2010, which committed Russia and the U.S. to reduce their deployed warheads to 1,550, deployed and non-deployed launchers to 800 by February 2018.[59] New START requires reductions in U.S. ICBMs from 450 to 400, all to be armed with one warhead after the 2020 NPR, and SLBMs will decline from 288 to 228.[60] Then in a 23 June 2013 speech at the Brandenburg Gate in Berlin, Obama called for further nuclear weapon reductions by "one third," or to about 1,000 warheads.[61] and Regarding tactical nuclear weapons, in the 2010 NPR the Administration decided to retire the U.S. Navy's nuclear-armed *Tomahawk* cruise missile (TLAM-N), possibly amounting to 157 missiles and over 330 warheads.[62] The U.S. is estimated to have about 500 tactical nuclear weapons.[63]

---

[58] From summary of Nuclear Posture Review, see, U.S. Department of Defense, "2010 Nuclear Posture Review (NPR) Fact Sheet," April 6, 2010, *Department of Defense*, <http://www.defense.gov/npr/docs/NPR%20FACT%20SHEET%20April%202010.pdf> ; for full report, see, Department of Defense, *Nuclear Posture Review Report, April 2010*, <http://www.defense.gov/npr/docs/2010%20nuclear%20posture%20review%20report.pdf>.

[59] Macon Phillips, "The New START Treaty and Protocol," *The White House Blog,* April 10, 2010, <http://www.whitehouse.gov/blog/2010/04/08/new-start-treaty-and-protocol>.

[60] "U.S. will cut Air Force nuke missile force by 50," *Associated Press,* April 8, 2014.

[61] The White House, "Office of the Press Secretary, Remarks by President Obama at the Brandenburg Gate --Pariser Platz, Brandenburg Gate, Berlin, Germany," June 19, 2013, *The White House*, <http://www.whitehouse.gov/the-press-office/2013/06/19/remarks-president-obama-brandenburg-gate-berlin-germany>.

[62] Jeffrey Lewis, "When the Navy Declassifies…," *Arms Control Wonk Blog,* July 12, 2012, <http://lewis.armscontrolwonk.com/archive/5499/when-the-navy-declassifies>; Hans M. Kristensen, "U.S. Navy Instruction Confirms Retirement of Nuclear Tomahawk Cruise Missile," *FAS Strategic Security Blog,* March 18, 2013, <http://blogs.fas.org/security/2013/03/tomahawk/>.

[63] "Nuclear Weapons: Who Has What At A Glance," *Arms Control Association,* Updated: November 2013, <http://www.armscontrol.org/factsheets/Nuclearweaponswhohaswhat>.

Unfortunately both Russia and China are not reducing the role of nuclear weapons in their strategies and they are building up their respective nuclear weapons capabilities, which could diminish Washington's ability to deter Moscow and Beijing from initiating limited conflict. Russian nuclear weapon modernization included the new multiple-warhead RS-24 *Yars* ICBM, the Project 955/955A *Borei* SSBN, an new rail-launched ICBM,[64] an intermediate range nuclear missile sometimes called the RS-26 *Frontier*[65] and a new strategic bomber with the program name *PAK-DA*.[66] Furthermore, Russia is estimated to have about 2,000-3,000 tactical nuclear weapons[67] to include a class of new small nuclear weapons to be fired by mobile artillery systems. Over the last decade major Russian military exercises of the *Zapad* series have included more sophisticated early use of tactical nuclear weapons. As already noted, China is expanding its ICBM and SLBM forces as it likely develops a new strategic bomber, and may have 650 tactical nuclear weapons. There should be little confidence in Chinese nuclear policies of "No First Use"; the PLA can be expected to use tactical nuclear weapons for intimidation and warfighting from the beginning of a conflict.

---

[64] "Project development BZHRK MIT," *MilitaryRussia.Ru Web Page,* April 23, 2013, <http://militaryrussia.ru/blog/topic-738.html>.

[65] "RS-26 Frontier/ KY-26 Vanguard," *MilitaryRussia.Ru Web Page,* July 5, 2011, <http://militaryrussia.ru/blog/topic-553.html>.

[66] "The financing of the PAK DA project has already been started," *Russian Aviation Web* Page, February 14, 2014, <http://www.ruaviation.com/news/2014/2/14/2180/>.

[67] For estimate of 2,000 deployed weapons, see, Oliver Meier and Simon Lun, "Trapped: NATO, Russia and the Problem of Tactical Nuclear Weapons," *Arms Control Association, Arms Control Today,* January/February 2014, <https://www.armscontrol.org/act/2014_01-02/Trapped-NATO-Russia-and-the-Problem-of -Tactical-Nuclear-Weapons>; In April 2011 Gary Samore, a Senior Director on the White House National Security Council said that Russia may have a "few thousand" tactical nuclear weapons whereas the U.S. only had a "few hundred." See, "Obama Adviser Gary Samore: 'The Ball Is Very Much in Tehran's Court,'" *Radio Free Europe / Radio Liberty*, April 14, 2011, <http://www.rferl.org/content/interview_samore_russia_iran_us_policy/3557326.html>.

Furthermore, North Korea's and Iran's imminent and Chinese-assisted nuclear missile capability also call into question the Obama Administration's continued commitment to policies promoting nuclear disarmament. Pyonyang's revelation in April 2012 of its KN-08 6,000km range liquid-fueled ICBM, seen on a Chinese-made China Aerospace Science and Industry Corporation (CASIC) Sanjiang Corporation 16-wheel transporter-erector-launcher (TEL),[68] has received a disturbingly minimal response from the Obama Administration.[69] China's willingness to sell this TEL raises legitimate questions over whether it may have also helped with its missile. In 1985 China abandoned a second-generation mobile liquid-fueled ICBM program called the DF-22;[70] could its plans have informed the KN-08?

3. Budgetary disarmament

In 2012 warnings emerged that declines in U.S. military spending would undermine the "pivot."[71] Though the August 2011 Budget Control Act requires the U.S. Department of Defense to implement spending reductions of $487 billion over ten years, in November 2011 President Obama

---

[68] Richard Fisher, "China's Strategic Assistance to North Korea's Nuclear Program," *The International Assessment and Strategy Center Web Page,* April 21, 2012, <http://www.strategycenter.net/research/pubID.278/pub_detail.asp>.

[69] Richard Fisher, "Obama's deadly inaction on North Korean missiles," *The Washington Times,* November 11, 2012, <http://www.washingtontimes.com/news/2012/nov/11/obamas-deadly-inaction-on-north-korean-nuclear-mis/>; "American weakness and Korean consequences," The Washington *Times,* April 10, 2013, <http://www.washingtontimes.com/news/2013/apr/10/fisher-american-weakness-and-korean-consequences/>.

[70] John Wilson Lewis and Hua Di, "China's Ballistic Missile Programs, Technologies, Strategies, Goals," *International Security,* Fall, Vol. 17, No. 2 (1992), p. 31.

[71] Dan Blumenthal and Michael Mazza, "Asia Needs A Larger US Defense Budget," *The Wall Street Journal,* July 5, 2012, <http://online.wsj.com/article/SB10001424052702304803104576425414030335604.html>

told the Australian Parliament that, "reductions in U.S. defense spending will not -- I repeat, will not -- come at the expense of the Asia Pacific."[72] Indeed, there are plans to shift 60 percent of the U.S. Navy to the Pacific over ten year, but budget realities also dictate that the number of U.S. Navy ships will not grow for the next five years and number of submarines required, about 48 or so, may not be sustainable.[73] DoD's weak fiscal condition is compounded by the guillotine of "sequestration," or the legal requirement of the August 2011 Budget Control Act that forced a $1 trillion reduction in U.S. government spending starting January 2013 to meet budget deficit reduction goals.

By early 2014 Sequestration-forced cuts were impacting current capability; an Assistant Secretary of Defense for Acquisition was stating publicly that the Rebalance to Asia "can't happen" due to budget pressures.[74] In February 2014 Defense Secretary Chuck Hagel listed proposed defense cuts from 2015 forward if Sequestrations reductions continued: retiring 320 A-10 ground attack aircraft; possible early retirement of one nuclear powered aircraft carrier; reduction of Littoral Combat Ships from 50 to 32; possible retirement of six cruisers; a two year delay the Navy's F-35C purchases; active Army forces may fall from 520,000 to 440,000 and a new Ground Combat Vehicle program will be cancelled.[75] Budget cuts may also see the termina-

---

[72]"Remarks By President Obama To The Australian Parliament," November 17, 2011, *The White House*,
　　<http://www.whitehouse.gov/the-press-office/2011/11/17/remarks-president-obama-australian-parliament>; This pledge had also just been made by Defense Secretary Leon Panetta, see, Adam Entous, "US Won't Cut Forces In Asia," *Wall Street Journal,* October 25, 2011,
　　<http://online.wsj.com/article/SB10001424052970204644504576650661091057424.html>.

[73] Otto Kreisher, "Navy Fleet Will Not Grow For Five Years: CNO," *AOL Defense Web Page,* February 7, 2012,
　　<http://defense.aol.com/2012/02/07/navy-fleet-will-not-hit-313-no-growth-for-5-years-cno/>.

[74] Fryer-Biggs, op-cit.

[75] Secretary of Defense Chuck Hagel, "FY 15 Budget Preview," February 24, 2014, *Department of Defense*, <http://www.defense.gov/speeches/speech.aspx?speechid=1831>.

tion of production of the *Tomahawk* cruise missile and the *Hellfire* anti-tank missile, both of which would see rapid expenditure in a major conflict, and a program to upgrade the radar of F-16 fighters. A U.S. Air Force officer was reported to have said, "We're burning the furniture to save the house."[76]

V. What the US Needs To Deter China Into the 2020s

While the Obama Administration has done a laudable job of beginning to craft an American strategy for countering China's growing military challenge, its success is threatened by inadequate funding stemming from an Administration priority on domestic programs. In addition, the strategies produced by the U.S. Department of Defense carry a large Asia-focus that China is countering by building toward a future global reach. Beijing can in the future "Pivot" to the Persian Gulf or Latin America to divert U.S. attention. But this is already being done by Russia's increasing militancy in Europe. Furthermore, at least on a political level, the Obama Administration does not appear to anticipate the potential for much greater Russian and Chinese strategic coordination or even cooperation against United States and other democracies.

Part of the answer to meeting these challenges would be acknowledgement that United States national priorities resulting in diminished defense capabilities are no longer sustainable. By allowing power balances to shift in favor of adversaries, Washington is in fact tempting China and Russia to instigate conflicts that could lead to larger and far more expensive wars. Nevertheless, it is possible for the United States to deter China and meet its eventual challenge beyond Asia. This will entail preserving much of the current U.S. force with some important additions. These would include:

---

[76] Robert F. Dorr, "US Air Force Confronts Harsh Reality," *Air International,* April 2014, p. 10.

1. New Missiles for U.S. Forces

While the recent RAND report criticizes missiles as not offering all the flexibility of bombers, they can be developed to enable a far greater assurance of destroying their target, and thus, would contribute greatly to deterring a spectrum of Chinese aggressive actions. Ballistic missiles may be more expensive than subsonic cruise missiles but their higher speed complicates interception and countermeasures while increasing destructive capability and are much less expensive than subsonic combat aircraft that are increasingly sophisticated integrated air defense systems (IADS). When placed on submarines, new U.S. intermediate or medium range ballistic missile would also be more secure than land-based aircraft.

The U.S. could develop a family of medium to short range anti-ship ballistic missiles (ASBMs), and sell them to its allies. Washington should oblige Russian suggestions that the 1987 Intermediate Range Nuclear Force Treaty is obsolete, in large part due to new Chinese missiles in this class. Large numbers of U.S. and allied ASBMs could "neutralize" China's Navy much as the PLA intends for its ASBMs to "neutralize" the navies of the United States and its allies. China may or may not approve of mutual assured naval destruction (MAND); without a global navy a "rising" superpower will rise more slowly, which may present a powerful incentive to pursue "rules" or even "control" of military behavior.

2. Regional network of long-range ground, sea, air and space-based sensors.

Washington should place a high priority on the creation of an Asian regional long-range sensor network that would provide network members a near real-time warning of broad Chinese military activity in order to allow regional governments to pursue individual or coalition defensive responses. The assurance that initial PLA attacks against sensors would be compensated even partially by other sensor network members, plus the greater intimate warning of early PLA moves, would help reduce the chances for PLA success and thus deter possible further aggressive action. Having access to a near real-time total picture of Chinese military actions could allow network members to reduce the risk of PLA deceptions that would exploit its increasing capacity for large-scale trans-regional military movements. Furthermore, the potential for such reactions to be near immediate and coordinated among states extending from Northeast Asia to the South Pacific would weigh more heavily on Chinese leaders and serve to counter Beijing's oft-employed "divide and conquer" tactics. Over time, the imposition of such a level of "transparency" on the range of Chinese military activities could lead to greater Chinese interest in regional confidence building measures that in turn may lead to interest in verifiable arms control regimes.

The U.S., Japan and Taiwan have ground-based very long-range phased-array radar positioned close enough to China to be able to monitor naval activity, plus air and missile activity, while the U.S., Japan and Australia have long-range sky wave Over-the-Horizon (OTH) radar, all with the potential ranges of one to several thousand kilometers. The Philippines, Malaysia, Indonesia have the option of purchasing less expensive surface-wave OTH systems or can offer to host jointly-manned long-range radar from the U.S. or other network members.

3. Rail Guns

Electromagnetic launch or rail guns would offer Taiwan the ability to finally the era of "missile terror" from China. A railgun uses electricity to accelerate a projectile and as long as it has access to a power source it can continue to fire "artillery rounds" until maintenance is required. But what reverses the cost-ratio in favor of the defender is the likelihood that an "artillery round" may cost tens of thousands of dollars compared to the estimated $.5 to $1 million cost of PLA SRBMs. An early rail gun is offered by the General Atomics *Blitzer*, which in early testing has fired an artillery size projectile up to speeds of Mach 5 (1,700 m/s), with a potential range of 100km.[77] On a destroyer-size ship the Blitzer could have a magazine of 1,000 rounds[78] and a round could contain up to 100 "pellets."[79] At a rate of fire of 10 rounds per minute[80] it is conceivable that a force of 20 land-based *Blitzers* on Taiwan could loft up to 40,000 potentially hypersonic (Mach 5 and above) speed pellets. As an anti-invasion defense, a force of 10 *Blitzers* could pose an unacceptable threat to a potential PLA force of hundreds of formal and informal invasion craft. An industry source told the author that assuming full funding, this railgun might be ready for production by 2015, though current plans may result in production by 2019.

In addition, the U.S. should place rail guns on nuclear powered cruise missile and nuclear powered attack submarines. In the Taiwan Strait, such rail gun equipped submarines would allow the U.S. to help degrade a Chinese missile barrage, which would help convince Beijing to terminate their attack as risks of failure would increase significantly.

---

[77] "'Blitzer' railgun already 'tactically relevant', boasts maker," *The Register,* December 15, 2010, <http://www.theregister.co.uk/2010/12/15/blitzer_trials/page2.html>.

[78] Lundquist, op-cit.

[79] One hundred pellets is a conservative estimate based on the author's viewing a potential round displayed by the Boeing Company at the April 2010 U.S. Navy League exhibition in Alexandria, Virginia.

[80] Grace Jean, Office of Naval Research, "With A Bang, Navy Begins Tests On Electromagnetic Railgun Launcher," *Office of Naval Research Web Page,* February 28, 2012, <http://www.onr.navy.mil/Media-Center/Press-Releases/2012/Electromagnetic-Railgun-B AE-Prototype-Launcher.aspx>; and earlier ONR release suggested a goal of 6-12 rounds a minute, see, Geoff S. Fein, Office of Naval Research, Corporate Strategic Communications, "Navy's Record-setting Test to Showcase Railgun's Military Relevance Friday At Dahlgren," December 8, 2010, <http://somd.com/news/headlines/2010/12931.shtml>.

4. Adequate numbers of F-35B short take-off and vertical landing (STOVL) fighters

Beset by program delays and now the largest U.S. weapons program at $395.7 billion for a planned purchase of 2,443, the F-35 Joint Strike Fighter is a key target for budget cutters. Compared to legacy 4th generation fighters in its class, like the F-16 and F/A-18, it does offer a clear advance in terms of bringing useful levels of stealth and a major increase in sensor capability to the battle. Its infrared warning sensors alone reportedly can detect missile launches up to 200 miles away, meaning it could assist missile defense missions.[81] But with the end of F-22 production, it is also the only remaining U.S. 5th gen tactical fighter option for the US Air Force, Navy and Marines, albeit one that has risen to an average $130-160 million in price, approaching that of the formerly "unaffordable" F-22.[82] However, the budget-vulnerable F-35B vertical or short take-off or landing (STOVL) version could make the greatest contribution to deterring the PLA as it could turn the U.S. Navy's 13 Landing Helicopter Dock (LHD) amphibious assault ships into capable aircraft carriers. The F-35B also offers key allies like Japan, South Korea and Australia the quickest path to acquiring their own naval air power. For Singapore, widely expected to buy the F-35B[83] and Taiwan,

---

[81] Robbin Laird, "F-35 Will 'Revolutionize' Combat Power In The Pacific," *AOL Defense Web Page,* December 22, 2011, <http://defense.aol.com/2011/12/22/f-35-will-revolutionize-air-combat-power-in-the-pacific/>.

[82] The higher number comes from the U.S. General Accounting Office; for a discussion of the F-35's cost increases and its ramifications, see Jon Lake, "How Much Does An F-35 Cost?," *Air Combat Monthly,* June 2012, pp. 22-25.

[83] Greg Waldron, "In Focus: Singapore steps up deterrent capabilities," *Flight International,* February 1, 2012, <http://www.flightglobal.com/news/articles/in-focus-singapore-steps-up-deterrent-capabilities-367582/>.

which has also signaled its interest,[84] the F-35B offers 5th generation performance plus tactical concealment advantages, as it could also be employed from the protective cover of U.S. naval formations with carriers or LHD size ships.

5. Rapidly Deployable Ground Combat Forces

While new missile and air combat systems may be most relevant to contesting maritime, air and space domains, the United States will still need to deploy dominant ground forces to support allies and friends. A constant decline in the numbers of forward deployed U.S. Army and Marine forces points to the need for a reconsideration of new medium-weight mechanized ground forces designed for rapid air transport. This was one of the goals of the ill-fated "Future Combat Systems" cancelled in 2009. There remains a requirement for a family of powerful and defended 30+ton combat vehicles that can be transported by C-17 size aircraft. The U.S. should also be developing a follow on for the aging C-5 transport.[85] Another casualty of the fall of the Future Combat Systems was the Army-Air Force Joint Heavy Lift program, which envisioned a 20-30 ton capable heavy vertical lift transports.[86] The latter are needed to be able to respond with overwhelming force to surprise PLA operations against disputed islands in the East and South China Seas. This should be considered for a joint development program with Japan.

6. Pursue Low Cost Asymmetrical Programs, Especially for Taiwan

The U.S. and its allies will also require new lower-cost but very capable weapons to meet the PLA's already substantial advantages in numbers of

---

[84] Gavin Phipps, "Taiwan mulling HF-2E deployment on new corvette, F-35 purchase," *Jane's Defence Weekly,* June 6, 2012, p. 39.

[85] For a Lockheed concept of a next generation airlifter, see Graham Warwick, "Shape Shift," *Aviation Week and Space Technology,* February 17, 2014, pp. 40-41.

[86] "China's Advanced Helicopter Concepts," op-cit.

weapons, which are also getting more capable. There should be a priority for developing new missile, aircraft and artillery systems. For example, at an estimated price of $3.4 million each,[87] South Korea could assemble a force of 500 ballistic and cruise missiles for the price of about 18 recently acquired Boeing F-15K fighter bombers with an average cost of about $96.7 million. At a price of about $1.8 million each,[88] Taiwan could acquire up to 1,000 ATACMS short-range ballistic missiles for the cost of about 14 new F-16C fighters it has sought to purchase, costing about $131 million each. This cost advantage in the face of respective "mass" threats is why South Korea and Taiwan have developed new land-attack missiles over the last decade.

Only until recently Washington has quietly opposed new long-range missile acquisition by Seoul and Taipei, but this policy should be reversed to include sale and co-development assistance for new missiles. China does not consider missile control to be in its interest and has continued to develop new regional missiles that threaten Taiwan, Japan, South Korea and the Philippines. Due to their cost efficiency, it is now necessary for Washington to help its friends acquire needed theater missile systems to deter China's missile forces.

There is also a real need for an effective $30-$40 million 4+ gen fighter due to the failure to contain costs for the F-35. Many countries are settling on the $40-$50 million Swedish SAAB *Grippen* to fill this requirement. The U.S. should use its new U.S. Air Force trainer program to also develop a new lightweight and inexpensive fighter. It need not be very stealthy to reduce

---

[87] "S. Korea to Build 500-600 More Missiles," *Chosun Ilbo,* May 22, 2012, <http://english.chosun.com/site/data/html_dir/2012/05/22/2012052200636.html>.

[88] Price derived from recently announced U.S. sale to Finland of 70 ATACMS for $132 million, see, "Finland orders M-39 Block 1A ATACMS from US," *www.army-technology.com web page,* June 7, 2012, <http://www.army-technology.com/news/newsfinland-orders-m-39-block-1a-atacms-united -states/>, This price, however, could be lower.

costs but should be armed with the latest weapons in order to meet the latest Chinese and Russian fighters. In addition, an upgraded version of Taiwan's Indigenous Defense Fighter (IDF) equipped with more powerful General Electric F404 or F414 turbofans would also meet this requirement.

The U.S. Navy is also developing a new class of highly accurate naval artillery shells, first to arm the new 155mm cannon on the new *Zumwalt* class cruiser and then the more widespread 127mm naval gun. To this competition Lockheed Martin has introduced a new winged artillery shell that it claims has a 120km range and will have sensors to enable attacks against moving targets.[89] This shell will also be comparable in cost to the $60-$70,000 *Hellfire* anti-tank missile, conveying a great cost advantage when attacking multi-million dollar PLA assault and transport ships. Taiwan's 600 155mm artillery systems potentially could use this precision-guided shell.

7. Reinvestment in new strategic and tactical nuclear weapons

The uncertainties regarding China's nuclear capabilities, Russia's modernizing nuclear forces and large tactical nuclear arsenal, and the onset of North Korean and Iranian long range nuclear strike requires a far more open assessment of nuclear threats for U.S. and allied policy makers. These threats also requires a new policies that halt further U.S. nuclear arms reductions, and that prepare for nuclear force modernization and growth if deemed necessary. It is furthermore necessary for Washington to try to assemble an international coalition to that offers China a choice: halt your proliferation of nuclear and missile weapon technology, come clean about past proliferation, enforce existing laws and regulations and begin a process that leads to far greater transparency and assurance, or the coalition is going to take defensive measures.

---

[89] Author interview, U.S. Navy League Exhibition, Washington, D.C., April 9, 2014.

The existence of a 5,000 km network of tunnels and the clear construction of extensive new tunneling to support existing and new PLA Second Artillery bases is enough to cast doubt on assumed public estimates of the number of PLA missiles and nuclear weapons. Even without considering General Yesin's estimates, and he should be invited to explain his methods and evidence, absent a full understanding of China's nuclear basing tunnels it is necessary to suspend nuclear reductions underway with Russia. In addition, the U.S. should prepare to replace reduced warheads on Minuteman ICBMs and Trident SLBMs and due consideration should be given to increasing the survivability of U.S. land-based ICBMs, as there should be full funding for the "SSBN-X" successor to the Trident SSBN and a new long-range bomber.

The Obama Administration's 2020 retirement of U.S. Navy Tomahawk TLAM-N was an act of unilateral disarmament and this capability must be replaced. There should be a crash program to develop a new nuclear armed *Tomahawk* or JASSM-ER based tactical cruise missile. The Administration's plan to rely on F-35 delivered tactical nuclear weapons is not as secure as placing tacnukes on very hard to find SSNs. There is a crucial requirement for this capability in Asia to deter China's use of its tactical nuclear forces, and to deter North Korea and Iranian use of their nuclear arsenals. These weapons are also needed to help deter China's use of conventional force when the U.S. cannot respond rapidly enough with conventional weapons.

8. Revived Moon Program

China and Russia may very well make the Moon the next military high-ground. From the Moon China can survey and target military Earth satellites or damage future space stations or future space-based solar energy collectors. Should they cooperate, Russia and China may be able to establish dual-use Moon bases more rapidly. In this light the Obama Administration's early 2010 cancellation of the Bush Administration's program to return to the

Moon was a strategic mistake. This can be recouped to some degree by Washington taking the lead in encouraging multi-government partial funding of a private commercial consortium to establish Moon facilities. This would allow the U.S. quickly counter Chinese or Russian military use of the Moon.

VI. Conclusion

After the failure of its attempt to forge a partnership with China in 2009, the Obama Administration's Pivot of 2011 was welcome in many quarters in Washington and among U.S. friends in Asia, as an overdue adjustment made necessary by China's increasing truculence, continued military buildup and proliferation of dangerous weapons to its rogue allies. This Pivot, however, is not secure. It is vulnerable to possible severe U.S. military spending reductions, as it also does not fully respond to the breadth and depth of the looming Chinese challenge. These include China's preparations for potential wars as it also increases the A2/AD capabilities of most concern for the Pivot, its continues proliferation of nuclear and missile technologies designed to strengthening of proxy allies, made more acute as China gathers power projection forces, could generate new Chinese Pivots to counter that of the United States, and finally, an apparent delinking of consideration of how China's potential for unclear breakout could severely undermine the Pivot. However, it is possible for Washington to rise to this challenge by considering new ways of coordinating existing military capabilities with allies and friends, considering new military capabilities and sharing them with friends and allies, while reconsidering old diplomatic and policy constraints that would prevent U.S. forces and those of its friends and allies from realizing these new capabilities.

China's Military Capabilities toward Taiwan: Use and Implication

Ming-Shih Shen

(Associate Professor, Graduate Institute of Strategic Studies,

National Defense University Taiwan)

Ⅰ. Introduction

Since the PLA reforming and modernizing its military, many experts agreed that the military balance shifting steadily in China's favor. As the "Defense Policy Paper Blue No.5", published by New Frontier Foundation, Democratic Progress Party (DPP), argued that People's Liberation Army(PLA)already had the operational capabilities of responding to Taiwan issues in 2007, and even surpassed Taiwan military power in 2010. In the near future, in 2020, PLA will continue to get decisive capabilities for a large-scale operation against Taiwan.[1]

Most of the articles on media or newspaper take the same point as "Blue Paper", to concern PLA's capabilities against Taiwan, because the PRC has continued to maintain an aggressive posture toward Taiwan. But the military capabilities of China cannot be seen equal to military threat directly. It depends on the intention and strategy beyond the military capabilities establishment.

Intention is an object or goal, it tells the outside world what the "decision-making black box" happen? Especially the PRC's intention of increas-

---

[1] New Frontier Foundation, "China's Military Threats against Taiwan in 2025," *Defense Policy Blue Paper NO. 5*, March 2014, p. 5.

ing and enlarging military capabilities refers to the purpose of the PLA strive for. On the official media of China, you will see many "old voice" from party leaders, to explain and emphasized the PLA's intention. Such as U.S think tank, Jamestown Fundation, looked at the intention of PLA, argued "...accelerate modernization of national defense and the armed forces so as to strengthen China's defense and military capabilities... should resolutely uphold China's sovereignty, security and territorial integrity, and ensure its peaceful development."[2] This article is critically to tease out the intention of PLA and the strategy behinds.

When Xi Jinping became the party General Secretary and the Chairman for the Central Military Commission, he also restated the familiar set of priorities for the military, which is "building a people's army that follows the Party's command, has the ability to win battles and has a fine work style."[3] Xi's request to the PLA is different from Jiang Zemin's Five Sentences and Hu Jintao's New Historic Missions, Xi asked PLA should have "China dream", "strong military dream" to be PLA's long term goal of military modernization. We can see that the PLA will be the key pillar of developing the China dream.

According to Foster's concept, strategy is a choice, it reflects a prefer strategy confronts adversaries, but some things simply remain beyond control or unforeseen.[4] Strategy provides the stats direction for the coercive or persuasive use of the power to achieve specified objectives. Objectives pro-

---

[2] Jamestown Foundation, "PLA Deputies Offer Clarifications of Military Intentions,"*China Brief*, Vol. 13, Issue 6 (2013), <http://www.refworld.org/docid/5146e4bd2.html>.

[3] Jamestown Foundation, "PLA Deputies Offer Clarifications of Military Intentions," ibid.

[4] Gregory D. Foster, "A Conceptual Foundation for a Theory of Strategy," *The Washington Quarterly*, Winter 1990, pp. 47-48.

vide purpose, focus and justification for the actions embedded in a strategy.[5] So the U.S. expert Art Lykke developed the theory of strategy to be a Model, to justify the strategy must balance ends, ways and means. If these are not balanced, the strategy would involve some dangerous risks.[6]

It is interesting to note that the rapid growth of China's military capabilities in recent years both in the quality and quantity will possess diverse and strong means to coerce Taiwan into negotiation and unification by the military capabilities. But this paper tried to evaluates the PLA's military capabilities form the framework of intention and strategy, to study PLA how to use these capabilities to be the threats and their implications to Taiwan.

II. PLA's capabilities toward Taiwan

China has never renounced military actions and preparations against Taiwan, and continues to formulate operational guidance and plans targeting Taiwan, even under the cross-strait relations rapprochement. In recent years, the Nanjing and the Guangzhou Military Regions have been outfitted with new major weapon systems and equipment and currently possess a variety of capabilities against Taiwan. The end of PLA military exercise in recent years is to rapidly end an island conflict and reduce the possibility of U.S. interference. Furthermore, China continues to protest against U.S. arms sale to Taiwan, demanding that U.S. sequentially reduce its arms sales to Taiwan each year until final termination. It hopes to hinder the further improvement of Taiwan defense capabilities and expand the gap in military power across the Taiwan Strait.[7] From strategic perspective, it mean that the China's final

---

[5] Gregory D. Foster, "A Conceptual Foundation for a Theory of Strategy," *The Washington Quarterly*, p. 50.

[6] Arthur F. Lykke Jr., "Toward an Understanding of Military Strategy," in *Military Strategy: Theory and Application* (Carlisle, PA: U.S. Army War College,1989), pp. 3-8.

[7] ROC MND, *2013 Quadrennial Defense Review*, 2013, p. 18.

ends toward Taiwan not change yet, but the ways and means become more flexible.

Taiwan also beware of this situation, in *ROC 2013 Quadrennial Defense Review*, it concluded that PLA's 6 military capabilities as follows:[8]

1. Joint intelligence, surveillance and reconnaissance capabilities

2. Strike capabilities of the PLA Second Artillery

3. Integrated air operation capabilities

4. Integrated maritime operation capabilities

5. Integrated land operation capabilities

6. Information and electronic operations capabilities

These 6 capabilities are regular military function and get more high-technological weapon systems or operated digitalized. But not noticed the integrated joint operation and asymmetric operational capabilities of PLA.

But from the perspective of DPP's *Defense Policy Blue Paper*, PLA's military capabilities will lead to the performance capacity of its weaponry greatly surpassing Taiwan in the near future, is already posing an enormous and palpable threat to Taiwan's defense security. The PLA military threats against Taiwan come from 4 capabilities include:[9]

---

[8] ROC MND, *2013 Quadrennial Defense Review*, pp. 18-22.

[9] New Frontier Foundation, "China's Military Threats against Taiwan in 2025," pp. 43-44.

1. Information and cyber warfare capability

The incursions against Taiwan's digital territory by China's cyberwarfare will endanger the functioning of Taiwan's society and the operations of the government, and could cause tangible damage to critical infrastructure, leading to loss of life or property.

2. Missiles precision strike capability

The growth in the number of Chinese missiles targeting Taiwan will gradually decelerate, but with the introduction of new types and models of missiles, the improvements to precision and diversification of warhead types will only increase, and will remain the first strike force that PLA forces will employ to coerce or to execute a swift and decisive victory to take over Taiwan.

3. Air and space operational capability

The expansive range of the PLA's air defense missiles has already embraced Taiwan within a de facto air defense identification zone, and when the 5th generation fighters enter into service by 2020, the PRC will achieve clear airpower superiority over Taiwan, and control of the airspace over Taiwan's territory will become increasingly tenuous, as Taiwan national defense strategy of credible deterrence cannot be executed, and resolute defense will be challenging to sustain.

4. Maritime operational capability

The PLA Navy is transforming to be modern, balanced fleets consist of surface, underwater, and carrier vessels, with the capacity to overpower Taiwan, confront Japan, and deter the United States, such that Taiwan issue will become a comfort zone new historical mission.

We can see the difference between the reports of ROC MND and DPP's Blue Paper, they outlined PLA's conventional capabilities only both, but not stated the key role of asymmetric operational capabilities. *ROC 2013 Quadrennial Defense Review* had stated China's "three-front war", the strategies of China's attempting to gain the right of military actions to prevent other countries from interfering with military conflict in the Taiwan Strait.[10]

China's military threat is characterized by asymmetric operational capabilities and conventional combat power. The asymmetric operational capabilities include strategic space operational power. Under in the information warfare, PLA is capable of exerting its electric-magnetic power. In acupuncture warfare, it is expected that the PLA can launch precision strikes against Taiwan by both cruise and guided missiles. Even beyond Taiwan, attack U.S. battle ships by ASBM. Finally, when it comes to the unlimited warfare, if China succeeds in integrating military, non-military and sur-military means, the armed forces of Taiwan will be under tremendous pressure in carrying out its military campaign.[11]

III. PLA's Intentions toward Taiwan

Taiwan's military security environment has been deeply influenced by China's military modernization. China has openly declared its intention to resolve the issue across the strait peacefully, but its relentless effort of military modernization un-nerves Taiwan and makes the situation filled with uncertainties. On the issue of Taiwan, China's main objectives are to force

---

[10] ROC MND, 2013 Quadrennial Defense Review, p. 18.

[11] Ming-Shih Shen& Chen-Tin Tsai, "The Factors Analysis of ROC Military Organization and Force Structure," *Taiwan Defense Affairs*, Vol. 4, No. 2 (Winter 2003/2004), p. 109.

Taiwan to follow the terms of "one state, two systems", peace talk, and resort for force when necessary or intervention by foreign powers.[12]

Therefore, when China aimed at coping with conflicts effectively and speedily before the international society response, relies on developing the so-called "assassin's mace" (shashoujian), setting the tempo and goals of operation in the process of carrying out the military action against Taiwan.

Under the control by the party, PLA officers enthusiastically support the defining objective of China's national strategy, which is to see China assume the status of a great power. All important decisions inside PLA are made by communist party committees that are dominated by political officers.[13] This system assures that the interests of the party's civilian and military leaders are merged, and for this reason new Chinese soldiers entering into the PLA swear their allegiance to the CCP, not to the PRC constitution or the people of China. It is therefore essential to view Chinese military strategy through the conceptual lens of CCP goals. Nationalism and the weight of the past are important factors. A strong China will never again be subject to the humiliations of the past. China's leaders believed that the key to great power status is to build a global-class economy and military. This requires maintaining a stable external environment to support high levels of economic growth.

Since China want to maintain a peaceful environment to develop a stable growing economy, the conflicts with neighbor countries is to be avoided. But the issues involved sovereignty like Taiwan, the East China Sea and the

---

[12] Ming-Shih Shen& Chen-Tin Tsai, "The Factors Analysis of ROC Military Organization and Force Structure," p. 132.

[13] U.S. DoD, *Military and Security Developments Involving the People's Republic of China 2013*, 2013, p. 24, <http://www.defense.gov/pubs/2013_China_Report_FINAL.pdf>; Ian Easton and Mark Stokes, *Half Lives: A Preliminary Assessment of China's Nuclear Warhead Life Extension and Safety Program* (Arlington, VA: Project 2049 Institute, July 2013), pp. 12-16, <http://project2049.net/half_lives_china_nuclear_warhead_program.pdf>.

South China Sea are the exceptions obviously. Then PLA leaders actively support China's present disputed islands strategies and policies. Military leaders feel that, in addition to serving national strategic objectives, the policies provide the best means of acquiring the capabilities required in high-technology warfare. Reconstituting the PLA into a modern military force has been the goal of the military modernization program the PLA has pursued since the early 1980s. Lack of information about the military modernization program, in turn, is also the source of much uncertainty about China's future intentions.[14] However, the growth of China's military power must be accompanied by greater clarity of its strategic intentions in order to avoid causing friction in the region.[15]

Andrew Scobell outlined that China's stated goal for Taiwan is clear: unification with China.[16] Since the late 1970s, China has proposed that this be accomplished peacefully through dialogue across the Taiwan Strait. China adopted a long-term horizon for unification and emphasized flexibility.[17] Despite China's more conciliatory policy toward Taiwan during the post-Mao era, China has still refused to renounce the use of force to achieve unification. China does see a key role for military force to push Taiwan move to the framework of unification.

In 2013, the U.S. think tank, 2049, explored China's strategic goals and describe the military measures that the PLA is implementing to achieve these goals. This study argued that theater geography, financial asymmetries, and

---

[14] Ron Montaparto, "China as a Military Power," *Strategic Foreum*, No.65 (December 1995), p. 3.

[15] U.S. DoD, *Sustaining U.S. Global Leadership: Priorities for 21st Century Deffense* (Washington D.C.: U.S. DoD, 2012), p.2.

[16] Andrew Scobell, "China's Military Threat to Taiwan in the 21st Century: Coercion or Capture?" *Taiwan Defense Affairs*, Vol. 4, No. 2 (Winter 2003/2004), p. 19.

[17] Andrew Scobell, "China's Military Threat to Taiwan in the 21st Century: Coercion or Capture?" p.19.

gaps in international law encourage the PLA to engage in a projectile-centric strategy that is destabilizing in nature, yet unlikely to change because it is the most effective means available for assuring that the Chinese threshold for victory could conceivably be reached in a conflict.[18] This paper argued the PLA will use it's military ways and non-military options to support China's end toward Taiwan.

It is confirmed by the China's new Defense White Paper, "Diversified Employment of China's Armed Forces," stated PLA should safeguarding national sovereignty, security and territorial integrity, and supporting the country's peaceful development.[19] There are several facts to prove PLA's intention. First fact is double-digit percentage increase of the PLA's budget. Although China's overall economic growth rate has declined, 2014 defense budget reflects the biggest increase in the three years and continues a several-decades-long trend of double-digit increases. Many experts assume that the real total is higher to meet the requirement of PLA arms leap advanced.[20]

PLA leaders placed the calls for "combat readiness" which had alarmed foreign observers amid Sino-Japanese tensions in the East China Sea within the PLA's focus on improving the quality of its soldiers. With China's aggressive approach to the region, PLA has sowed suspicion among its neighbors, who fear not only economic but military dominance. China is engaged in a dangerous dispute with Japan over the sovereignty of islands adminis-

---

[18] Ian Easton, "China's Military Strategy in The Asia-Pacific: Implications for Regional Stability," *Project 2049 Institute*, September 26, 2013, p. 3, <http://www.project2049.net/documents/China_Military_Strategy_Easton.pdf>.

[19] Ministry of National Defense of the People's Public of China, "Diversified Employment of China's Armed Forces," April 2013, <http://eng.chinamil.com.cn/special-reports/2013-04/16/content_5301690.htm >.

[20] "China's Disturbing Defense Budget," *The New York Times*, March 9, 2014, <http://www.nytimes.com/2014/03/10/opinion/chinas-disturbing-defense-budget.html?_r=0>.

tered by Japan in the East China Sea, raising fears that frequent movements around the islands by Chinese military patrols and Japanese fishing vessels could spark a conflict.[21] In November 2013, China stunned Japan, Taiwan, South Korea and the United States by declaring a new air defense zone over parts of that sea. This official statement of ADIZ on East China Sea rised the tension among U.S., China and Japan.

Finally, PLA is going global and faces a new set of challenges as it adjusts to the protection of Chinese interests overseas. None of these are new things, but the little variations upon official themes still warrant some attention. China has also been intimidating Southeast Asian nations that oppose its territorial claims in the South China Sea, with its fisheries and reputed oil and gas reserves. China seems intent on controlling the entire South China Sea, an important shipping route for energy resources from the Middle East. It is locked in territorial disputes in the area with Vietnam and other ASEAN members.[22] While some experts predict that China is investing in new systems, including submarines, surface ships and anti-ship ballistic missiles, that could be used to further intimidate neighbors or deny the United States access to Asian waters to defend its allies.

There is also one especially unpleasant fact here. Relatively small Taiwan almost certainly cannot cope with much larger China with respect to military strength, and the relative disadvantage seems certain to increase with the passage of time. In the scenarios of future military conflict on Taiwan Straits, China has the people, the growing economy, the technological advances, and the general wherewithal to prevail, whether quickly or even-

---

[21] ibid.

[22] Hiroshi Murayama, "Has China's Military Expansion Peaked?" *Asian Review*, March 30, 2014.

tually.[23] Of course, Taiwan hopes to have U.S. support to deter China, to defend against PLA attack operation, it depending on how some "final crisis" might develop.

IV. PLA's Implication to Taiwan

If PLA plan to invade Taiwan, a quick victory is the preferred option so that foreign forces would have minimal response time to show up to support Taiwan's defense. For China, if a war were considered absolutely necessary, then a quick operation or campaign would be best choice. China's military has serious limits on its power projection capabilities before 2009. PLA was characterized as having "short arms and slow legs."[24] But according to Scobell, PLA's power projection abilities continue to improve and while a Taiwan operation would undoubtedly stretch these, such an operation would not necessarily prove a bridge too far. PLA should have "long arms and fast legs," these capabilities make its power projection beyond Taiwan to the first island chains. In any invade Taiwan operation, PLA will try to keep operations familiar and closed to the comfort zone of the troops.[25] If Taiwan combat readiness and weaponry system still follow the old fighting concept and training, that will provide a comfort zone on Taiwan Strait to PLA.

For the defense establishment in Taiwan, caution is therefore needed in that China has been emphatic of an integrated strategy in the case related to Taiwan's sovereignty. Backed by "verbal initiatives" and "physical threat",

---

[23] Eric A. McVadon, "Arming Taiwan for the Future: Prospects and Problems." *Taiwan Defense Affairs*, Vol. 4, No. 2 (Winter 2003/2004), p. 154.

[24] Russell D. Howard, *The Chinese People's Liberation Army: 'Short Arms and Slow Legs'*, *INSS Occasional paper No. 28* (CO: U.S. Air Force Academy Institute for National Security Studies,1999).

[25] Andrew Scobell, "China's Military Threat to Taiwan in twenty First Century," in Martin Edmonds & Michael M. Tsai, eds., *Taiwan Defense Reform* (London: Routledge, 2006), p. 34.

China's approach to Taiwan not only underlines sufficient military preparation, but also stresses combined effort of politics, economics, diplomacy and culture.[26] Here, "verbal initiatives" refer to successive efforts to peacefully resolve issues related to Taiwan. These include effort to convince the Taiwanese people to accept the political agreement toward unification and an international environment conducive to unification of China or at least one against Taiwan's independence.

"Physical threat" refers to an unrelenting position not to renounce force to settle issues related to Taiwan. To demonstrate Beijing's resolve, PLA is making sustaining effort both to strengthen its ability to win a local war under information-technology conditions, and to deter Taiwan by large-scale defense modernization. As Beijing has been emphatic, there is no guarantee to resolution of the issues related to Taiwan's sovereignty and strengthening military capabilities and preparing for military operation seem to be the bottom line. While "physical threat" leaves much to anticipate, "verbal initiatives" promoted by the "harmonious world" theory declared earlier by Hu Jintao, also followed by Xi Jinping, have already borne fruit.

In fact, before 2008, with closer contact between Beijing and Washington, and as a result of Beijing's approach to "dominating Taiwan via the US", China has extended the discourse of "verbal initiatives" from cross-Straits area into the overall Asia-Pacific region.[27]

It also has to be noted that while the basic nature of the strategic terms—looking for softer approaches with soft power, and harder approaches with hard power—remains, Beijing has been more flexible in forms and methods. Beijing today, in its dealing with international issues, begins to

---

[26] An-Kang Hu, *China's Grand Strategy* (Hunchao: Chikang People's Publishing House, 2003), p. 116.

[27] Sao Nin ed., *2006-2008 Taiwan Strait Situation Report* (Beijing: Joechou Publishing House, 2006), pp. 193-206.

pursue an integrated strategy that is characterized by being softer by using soft power but with hard power behind. In other words, Beijing takes it as an ideal objective as it is preparing itself for the following objectives: Taiwan accepts the political conditions or that Taiwan at least refrains from officially declaring Taiwan's independence. These strategic objectives are actually highlighted by all the top leaders throughout the history of PRC. It is an ideal objective as it means an approach to unification of China with Taiwan but without resort to war. Taiwan, under this potential tendency of "verbal initiates" and "physical threat" shaped by Beijing's objective and moves, obviously bodes ill. Facing the threat either tangible or intangible, Taiwan is therefore highly advised to map out effective strategies to defend the homeland, among which the key is building military capabilities sufficient for homeland security.

Shambaugh had written an article in 2000, stated China's rapid strides in offensive capabilities in 1990s, given Taiwan's formidable natural and military defenses. But the strategic balance is shifting. He outlined in the next decade Taiwan will be negotiating with China from a position of inferiority.[28] We can see the tendency is coming true right now.

PLA has recently been of the view that amphibious landing operation in hi-tech conditions needs air supremacy, without which there is no sea supremacy either. It has to be pointed out this understanding is an important conceptual breakthrough among the PLA leaderships. PLA in the past admitted that there were relative disadvantages to the PLAAF when it came to some cross-Straits military conflicts. Air supremacy sought after by the PLAAF could be undercut by factors such as fewer airfields at the frontal lines along the southeast coasts of China, insufficient high-tech fighters in

---

[28] David Shambaugh, "A Matter of Time: Taiwan's Eroding Military Advantage," *The Washington Quarterly*, Vol. 23, Issue 2 (2000), pp. 119-133.

readiness, outdated sea-carried platforms, limited tracking distance of radar and a lack of mid-range missile in service.

These concerns however have been unremittingly addressed by PLA with the rise of economic strength. It proves that with a giant march of PLA's combat capabilities, air supremacy has been significantly enhanced to assure successful amphibious operation now. For instance, as Nanjng Military Region began to deploy J-10 and Su-27 fighters, the air supremacy held by Taiwan for years has been giving way to PLA leading to an imbalance of power between China and Taiwan.[29]

PLA is also keen to develop aircraft carriers for its expansion into maritime activities. Ever since 2011 when China launched its first plane carrier for a maiden run, there had succeed in trials on sea, and deployed in North sea Fleet.[30] It is anticipated that both the East Sea fleet and South Sea fleet will set up a carrier battle group as a means to strengthen sea power in the West Pacific and the Indian Ocean respectively. Between these two fleets, the former force structure is even more relevant to war across the Taiwan Straits. The East Sea fleet, in addition to countering an intervening US-Japan operation staged in the West Pacific, has capabilities to blockade or attack the eastern coast of Taiwan. The effect of the maneuver from the east coast of Taiwan could constitute a potential for either two-fold, several-fold battle fronts, or surrounding posture against Taiwan, or cutting short the strategic depth of Taiwan,[31] pressurizing air power and sea power of the Taiwan islands.

---

[29] U.S. DoD, *Military and Security Developments Involving the People's Republic of China 2012*, pp. 5-10.

[30] Ming-Shih Shen, "The Intention and Strategic Implications of the PRC's First Aircraft Carrier Trial Voyage," *Prospect and Exploration (New Taipei City)*, Vol. 9, No. 9 (2011), pp. 24-28.

[31] Ming-Shih Shen, "PLAN Sea Parade: Intention and Implication," *Strategic and Security Analyses (Taipei)*, No. 49 (May 2009), p. 29.

Some scholars, starting their observations from the above premises, proceed with the issues of joint operation in the self-defense effort of Taiwan but ending up with two different versions of understanding worth further discussions. To begin with, both schools of thought agree that air force and navy are important tools to deter and counter aggression in peacetime and early phases of wartime. However, given the fact that Taiwan's air force and naval components is significantly outnumbered by those of PLA, there exists a skeptical viewpoint that a saturation attack or feign air intrusion from across the Taiwan Straits by PLAAF's relatively inferior-quality fighters may effectively exhaust Taiwan's fighters and anti-air missiles, subsequently puncturing apart the overall homeland defense on Taiwan proper.

In the ensuring phases of war across the Taiwan Straits, as the second school concerning self-defense enterprises argues, PLAAF's updated fighters will find it relatively easier to carry out close-support missions for the amphibious landing of PLA's task forces. Based upon this reasoning and according to the second school, Taiwan proper is essentially where a decisive battle takes place, and ground battle is the key to final result of the war across Taiwan. There is, as the second school of thought goes on, no need to invest a proportionately larger share of defense budget in arms that may be not too much relevant to the combat utilities on the ground. The defense establishment of Taiwan instead should underline the covering and sheltering bulwarks on the grounds against the first strike from PLA. Taiwan should never underestimate itself or lose the fighting will to defend against China.[32]

The above two schools of thought will bring about different orientation of arms build-up in Taiwan and therefore cannot be seen as trivial. Both positions to be honest cannot be said baseless. However, they are not exclusive with each other but can accommodate the strengths of the other. To be sure,

---

[32] Chien-Hung Liu, "China's Growing Military Power and Its Implications," *Taiwan Defense Affairs*, Vol. 5, No. 3 (Spring 2005), p. 85.

the purpose of building air force and naval power is not reactively for pro-tracted war.

To deter PLA's intention to launch war against Taiwan, air force and naval power have to be identified as important means to raise Beijing's cost of invasion whether it is an act of blockade or direct attack. Without air su-premacy or being disadvantaged in maritime power across the Straits, Tai-wan cannot force Beijing to re-consider the ill consequences of a possible war as similar as those in the past history. That mean the issue of number may not be the key to get the military win. Although Taiwan has been out-numbered by China with respect to air/naval power, perhaps qualitatively and quantitatively, what counts as decisive to final victory lies at the tactics and operations levels. It is generally understood that flexible tactics and op-erations as well as manipulation of asymmetric warfare often dash the mo-rale of the opposite and prevail the overwhelming enemy, resulting in final victory of war.[33]

V. Taiwan's Strategic Options

Taiwan's defense strategy, epistemologically speaking, can be roughly divided into four levels—international, national, defense and military. At the international level, analysts explore the impact of instabilities in the Asia Pacific region on the cross-Straits relations and hence regional security. Field observation more often than not focus their attention on the followings: first, development of the cross-Straits relations; second, shift of balance power across the Taiwan Straits; third, likely responses of the neighboring countries as the result of the above changes; fourth, if the neighboring countries take military option to respond to the above changes, what could be the potential

---

[33] ROC MND, "The R.O.C.'s Defense Transformation: Building a Volunteer Force and Inno-vative & Asymmetric Capabilities," *Defense Security Brief*, Vol. 4, Issue 1 (January 2014), p. 3.

impact on Taiwan's self-defense strategy? Fifth, will Beijing's anti-access strategy, a response that have been much discussed by the observers, cause uncertainties in the US and Japanese calculation of their military security? However, most of the countermeasures the U.S. and its allies are taking in reaction to China's projectile-centric strategy are not sufficient. There is much more that needs to be done to assure that the U.S. will remain capable of projecting power into the Western Pacific in the face of China's offensive strike capabilities.[34]

Will the certainty be serious enough to the extent that the US and Japan take a stand-by attitude, given the conflict across the Straits escalates into war? And last but surely not the least, what strategy should Taiwan take to respond to the above situations that cannot be lightly dismissed? How does Taiwan deal with the conditions when there is no international intervention by military force and when self-defense has to be performed independently by Taiwan? Given international intervention does take place, how does Taiwan accommodate or coordinate with the rescuing foreign troops? All of these puzzles mentioned above—by no means exhaustive, of course, have to be put into the equation of Taiwan's strategic formula.

With respect to the national strategy, one very strong line of thought is that "comprehensive security" calls for the need to expand traditional pattern of thinking on the issue of security. To be substantial, national defense strategy in the past focused solely on factors within the realm of defense and military strategy. However, with the dynamic impacts of globalization and continual easing of relations across the Straits, national defense strategy needs to, first, integrate factors other than the conventional framework of national defense and, second, decide priorities among courses of action. One important

---

[34] Ian Easton, "China's Military Strategy in The Asia-Pacific: Implications for Regional Stability," *Project 2049 Institute*, September 26, 2013, p. 16, <http://www.project2049.net/documents/China_Military_Strategy_Easton.pdf>.

question in need of further exploration: what is the role of national defense strategy to address the current cross-Straits relations?

It seems that one of the issues to challenge Taiwan's defense establishment is how to cope with the imbalance and impact as a result of adjusting priorities of the national defense strategic objectives. It has been pointed out that Taiwan may be in lack of a national defense strategy. It also has been mentioned that Taiwan is in short of a clear strategic guideline, which leads to an obviously rudderless national defense strategy.[35] Another comment has it that since military tension across the Taiwan Straits has already receded national defense strategy should refrain from being too much provocative lest the current harmonious atmosphere be disturbed. Given the comments above, it is essential for national defense strategy as well as mechanisms to have a proper extent of supervision from above and integration horizontally. This is especially true that, with the improvement of cross-Straits relations but a rise of PLA's might, Taiwan's defense strategy needs to address both harmonious interaction with Beijing and defensive bulwark against PLA.

It stands to reason that major discussions of national defense strategy center upon the means to integrate various bureaucratic units under the Executive Yuan. It is expected that through integration of the branch units organizational functions and departmental resources can be effectively consolidated in support of defense strategy at the national level. For instance, in the wake of the Three Direct Links coming into effect, an ample scope of cooperation across the Straits on maritime security and joint strikes against criminal acts does come to light. As the result of increasing interaction, no longer

---

[35] Jean-Pierre Cabestan, "Taiwan's Strategic Goals and Action Plan," paper presented at "A New Strategy for a New Era:Revisiting Taiwan's National Security Strategy" (Taipei: The Center for Security Studies (MCSS), National Chengchi University &The Centre for International Security studies, University of Sydney, 27-28 August, 2011),
<http://www.mcsstw.org/web/content.php?PID=5&Nid=959>.

do the issues of security belong to the "territory"—functionally speaking—of the Ministry of Defense.[36]

It makes sense to argue that responsibility of security these days has to be shared among Ministry of Defense, Ministry of Transportation and Communication, Mainland Affairs Council, international sea/air ports, customs offices, Bureau of Consular Affairs, etc.. The issue of integrating bureaucratic functions will deliver impact on defense-related security and consequently change the nature of self-defense strategy. Under the current cross-Straits situation that is characterized by dynamic development loaded with uncertainties, a shift of self-defense strategy needs to be cautious. Factors that ought to be placed into the strategic equation include potential risks as a result of environmental changes and risk-prevention as well as risk-aversion measures beforehand.[37]

The pivotal point here is that there is no significant change of military strategy in terms of basic nature and major orientation. First of all, Beijing has not renounced military means to deal with issues related to Taiwan. Second and more importantly, PLA's combat readiness have already surpassed the need to take over Taiwan if the former determines to invade unilaterally. Thirdly and evidently, China has been making tremendous effort—combat readiness, strategic maneuver, field training—to prepare for power projection into the "first island chain" in the Pacific.

In light of the above, it would suffice simply to indicate that Taiwan self-defense strategy at the military level can be the most unpredictable. It is unpredictable especially when there is a short of military information—worse still is the moment before potential attacks when the enemy

---

[36] Ming-Shih Shen, "Taiwan Defense Strategy: Island Battle Perspective," *Strategic and Security Analyses*, No. 44 (2008), pp. 41-42.

[37] ibid.

seizes the initiative in the battle field. Taiwan defense strategy at the military-strategy level is therefore advisable to lay emphasis on PLA's *modus operandi* to invade Taiwan, Taiwan's operation guidelines and rules of engagement. Being broad in Taiwan self-defense strategy at the military level is also essential. If combat scenarios cannot be sufficiently comprehensive, or if threat and invasion from PLA have been constrained at rather limited and specific conditions, misjudgment will highly likely to occur, military strategy distorted, and troop deployment ineffective.[38]

As a matter of fact, any rational discussion on the self-defense strategy of Taiwan has to return to the fundamental. That is: The battle fields as a result of military clashes definitely taking place on Taiwan proper and the neighborhood of Taiwan Straits. Even if missile warfare, information warfare, or warfare beyond bounds conducted from areas remote from Taiwan may be one of options of PLA's tactical means, it is not a necessary factor leading to victory over Taiwan. In other words, to enter Taiwan proper and control the situation in hand, ground troops, whether they are PLA or not, are indispensable. Arguably as a result, the core of overall Taiwan self-defense strategy seems to be ground battle.

It should also be remembered that Beijing takes "active defense" as its strategic guideline. Under this guideline, Beijing seeks to seize strategic initiative by building a peaceful image, while actively makes preparation for military struggle in the future. To respond to Beijing's military strategy featured by "active defense", Taiwan needs to be engaged in "agile defense", checking PLA's "active defense". Emphasis has to be made by Taiwan that all the effort of arms build-up and purchases of advanced weaponry has nothing to do with offensive endeavor but defense-oriented. As long as Beijing renounces its force against Taiwan, there is certainly no need to further strengthen defensive measurements. Having argued in this way, however, the

---

[38] ibid.

bottom line is that when mentioning a defensive nature of defense, Taiwan should not restrict its strategic imagination and solely stick itself to the concepts of "being reactive", combat power preservation, and homeland operation.[39]

VI. Conclusions

In the course of international conflicts and military clashes, it is undeniable that force- the military capabilities- is inevitably the decisive bargaining chip as well as final resort. A state may risk a danger of a deteriorating force, begging the potential problems of a giant neighbor's salivated lust for invasion and inking of unequal treaties. Take Russian invasion into Georgia for instance. After the declared ceasefire between the two sides, Russia did not call a halt to its aggressive behavior toward Georgia. It instead speeded up its deployment in the Georgian territory, creating an atmosphere favorable to Russia in the latter phase of bargaining. This is an evidence case to show that military means is often the threshold to reaping more strategic interests.

To conclude, this paper is of the view that the defense establishment in Taiwan ought to actively strengthen its preparatory measures for self-defense operation so that Taiwan is able to cope with the threats from an ascending power of PLA. To meet the mission requirements, Taiwan has to continue the weaponry acquisition, aimed at high-tech military readiness. Taiwan also has to base up the current force structure and plan for defense operation against the physical intimidation across the Straits. More importantly, it is recommended that better force disposition and pragmatic troops training be emphasized in a series of joint operation preparations. Among them, annually held large-scale military exercises can surely elevate combat power, as it assumes that when the government and people in Taiwan have better trust in

---

[39] ibid.

their future stability, and they feel less vulnerable to the PLA, healthy relations across the Taiwan Straits can be sustained as well. Do not providing an excuse to use force by avoiding provocative actions that China could use to justify an attack.[40]

PLA may prove to be undeterable for its developing capabilities. Nevertheless, Taiwan must be prepared to demonstrate that any PLA attack will meet with immediate, stiff and spirited resistance.[41] In order to attain this goal, Taiwan should displaying very strong combat readiness, as evident through periodic Han Kuang military exercises. The series exercises were designed to counter the kinds of military threat from PLA.

In future, facing the military threat from PLA's military buildup cannot to afford to fail, and Taiwan must prepare defense under the guidance of defense strategy and strategic concept of "resolute defense and credible deterrence," and to achieve the goal of strategic sustainment and tactical decisiveness to defeat the enemy's intention to get rapid victory.[42] Taiwan military also need to establish more advance fighting capabilities that can meet joint operations requirements, and to achieve the objective of defense operations. For Taiwan, the best way of dissuading the PLA is shows of Taiwan's own military capabilities, and determination without being unduly provocative, and the possibility of outside military intervention alive.[43]

---

[40] Phillip Saunders, "Defending Taiwan: The QDR and Beyond," *Defense Security Brief*, Vol. 4, Issue 1 (January 2014), p. 3.

[41] Andrew Scobell, "China's Military Threat to Taiwan in the 21st Century: Coercion or Capture?" *Taiwan Defense Affairs*, Vol. 4, No. 2 (Winter 2003/2004), p. 32.

[42] ROC MND, *2009 Quadrennial Defense Review*, 2009, p. 116.

[43] Phillip Saunders, "Defending Taiwan: The QDR and Beyond," p. 3.